THE WELL-PLANNED UNIVERSE

EXPLORING THE TRUE NATURE OF OUR WORLD AND THE UNIVERSE BEYOND

Kenneth G. Davis

THE WELL-PLANNED UNIVERSE

Exploring The True Nature Of Our World

And The Universe Beyond

Published by Kenneth G. Davis

Printed in the United States of America

In memory of my wife of 45 years, Lin Marie, who provided unwavering support, and encouraged me to pursue writing this book.

TABLE OF CONTENTS

CHAPTER ONE

FROM CURIOSITY TO CLARITY

"The important thing is not to stop questioning. Curiosity has its own reason for existence. One cannot help but be in awe when he contemplates the mysteries of eternity, of life, of the marvelous structure of reality."
Albert Einstein, physicist

Fundamental Questions

"Ouch, he bit me." I was eight years old, sitting too close to a teeming ant hill. Those little red creatures were frantic and angry, and I soon learned to respect their territory. My curiosity piqued by observing the wonders of the world of insects, I began my journey of discovery.

Over time, my fascination with insect behavior developed into an interest in all living things. Eventually, this interest further broadened into a desire to learn as much as I could about the makeup of the world, its inhabitants, and its relationship to the rest of the universe. This curiosity about the nature of reality and the world in which we find ourselves lead me to pursue a degree in physics. Following university graduation, I went to work as a geophysicist in the petroleum industry, applying computer technology to the analysis of seismic data.

In the past few decades, scientific discoveries and the development of digital computer systems have led to a much-improved understanding of the physical properties of our world and the universe. Research into the characteristics and behavior of living things has also provided insight into why we find the world to be the way that it is.

My goal in writing this book was to use recent evidence from these diverse areas of study to find answers to fundamental questions about reality and the universe, including:

- What is the universe really made of?
- Why is the universe structured and organized the way it is?
- Where did everything that we can see, hear, and feel come from?
- Why does the universe exist at all?

Throughout history, philosophers have given much thought to these questions. With recent developments in science and technology, an opportunity now exists, as never before, to obtain answers. We no longer have to rely on the statements or beliefs of those who claim to have answers, but have only opinions.

Correct answers result from a proper investigation and require an objective look at all available relevant information. As you will discover, strictly following the evidence and using a multidisciplinary approach will provide answers.

The evidence required comes primarily from three sources:

- Scientific data from the physical world.
  Recent scientific discoveries have led to a new, deeper understanding of the universe. The physical laws and properties of the universe are now understood in much greater detail than ever before. Evidence from physics, astronomy, chemistry, and biology is particularly valuable. The special nature of the physical laws and properties of the universe lead to some remarkable conclusions.

- The behavior and characteristics of the many forms of life on Earth.
  The attributes and variety of life on Earth help to further explain and understand the significance of recent scientific discoveries.

- The development of digital computer systems.
  Recent developments in computers provide a deeper understanding of living things and how they interact with each other.

Accumulated scientific evidence from these areas leads to some amazing conclusions about reality and the universe.

Following the Evidence

I like to think of myself as an inquiring skeptic. I will not be convinced simply because an idea is widely published, sounds good, seems to make sense, becomes popular, or is promoted by people that I admire. As a natural-born skeptic, I require sufficient evidence before accepting or rejecting any claim or theory.

The proper use of evidence is more important than ever before, due to widespread misleading, deceptive, or outright false information. Disinformation, altered photos, and opinions disguised as facts are becoming all too common.

To be absolutely certain of something means that it is known beyond all doubt, and there is no need for a discussion of evidence. For example, there is no doubt that when a fire begins to spread in a forest, there will be smoke and flames as a result. There is no forest fire on Earth that does not produce smoke and flames. Likewise, when a basketball is tossed onto the surface of a standard basketball court, there is no doubt that it will rebound. Everyone accepts that the ball will bounce off the surface of the court. However, the details of precisely how fast or at what angle it will move after hitting the ground may lead to some discussion—as you may remember from high school physics class.

Few things in life are as certain as these examples. Nearly every conclusion arrived at in life involves a degree of uncertainty, so using evidence is essential. Understanding the proper use of evidence will make it possible to arrive at a correct conclusion.

Evidence consists of the information that helps to form a conclusion or judgment. Decision-makers, such as judges and juries, make important decisions every day based on the evaluation of evidence. In the most important situations, a decision will be considered valid only if it can be supported by the available evidence—beyond any reasonable doubt. "Beyond a reasonable doubt" is the highest standard of proof used in the United States and many other justice systems. It refers to a level of confidence such that there could be no reasonable doubt in the mind of a person using sound judgment.

The evidence presented to a judge and jury must be both material and logically relevant. By definition, “material evidence” has a significant relationship to the facts or issues under consideration. The “United States Federal Rules of Evidence” defines logically relevant evidence as “when evidence introduced has any tendency to make a material fact more probable or less probable than it would be without the evidence [1].”

A skilled investigator will take note of things observed, but just as importantly will record anything they expect to see, but don't. They will look for inconsistencies in what they observe, and for things missing or out of place at the scene of the investigation. They will ask, “Why do I observe this?” But they will also ask, “Why do I not observe that?” and, “What is wrong with this picture?”

Types of Evidence

Five main types of evidence are used in a court of law. They are all useful for investigating the fundamental nature of reality.

- ***Physical evidence*** refers to a tangible item with physical characteristics that tends to prove a material fact. It can be a physical substance or traces left behind from a relevant event, such as footprints that can be photographed and measured.

- ***Documentary evidence*** refers to some type of document, such as email, computer file, or magazine article, that supports a claim or assertion. It may simply consist of a written description of an event from an eyewitness.

- ***Testimonial evidence*** refers to a statement that tends to prove a material fact. An expert witness with special knowledge of a certain subject area sometimes gives testimonial evidence. It may also be presented by a layperson who has pertinent information.

- ***Demonstrative evidence*** comes in the form of a representation of something material, such as a diagram, map, drawing, graph, animation, simulation, or model. Computer systems have contributed greatly to the usefulness of demonstrative evidence.

- ***Scientific evidence*** comes from either experiments or observations. The results from scientific evidence are sometimes presented in the form of demonstrative evidence.

Each of the five types of evidence may be either direct or indirect evidence, or may have aspects of each.

Direct evidence supports the truth of an assertion directly, so it does not require inference on the part of a judge or jury. Eyewitness testimony, physical evidence, and documentation relevant to the issue are usually considered direct evidence. The usefulness of direct evidence depends upon the reliability of the source of the information.

Indirect—or circumstantial—evidence allows for an inference to be drawn that tends to prove or disprove a relevant point. Imagine a situation where, before going to bed one night, you notice a fresh blanket of snow on your lawn. The next morning you see footprints in the snow. The presence of footprints allows you to infer that someone has walked across the lawn while you were sleeping, even though you did not see anyone in your yard.

Establishing the truth requires more than just one isolated piece of evidence. Confidence in a conclusion increases substantially when it is supported by different types of evidence from multiple sources. Inferring something from a single piece of information can lead to an incorrect conclusion, particularly in the case of indirect evidence that may be interpreted in more than one way.

For example, consider a mother who has baked a fresh blueberry pie and put it on the counter to cool. After coming back from running an errand, she finds a piece of the pie missing. Her six-year-old son enters the kitchen with blueberry juice on his face. With these two pieces of evidence she can infer that her son had sampled the pie, even though she did not personally see him eat any. However, if the mother had seen her son with blueberry juice on his face before she entered the house, she should not conclude, with that single piece of evidence, that he had eaten some of her pie. She should first check in the house to determine if some pie is missing. Unbeknownst to her, her son may have been given a piece of pie by a neighbor (the neighbor having also made a pie since blueberries were in season at that time of year).

This clearly illustrates the importance of evaluating all available relevant information before coming to a conclusion. Making a decision without using all the available facts is sometimes referred to as "jumping to conclusions." Taking time to gather pertinent facts and data will ultimately result in more clarity, leading to greater confidence in conclusions and in the decisions made based on those conclusions.

A Mountain Meadow Discovery

Consider now this example of how various pieces of evidence come together to lead to a convincing conclusion. A group of hikers come across a remote, isolated mountain meadow. They observe wildflowers and other small plants randomly scattered throughout the meadow. They also discover a section of this meadow that has a number of neat parallel rows made up of one kind of plant, but with very few of the native plants or wildflowers seen elsewhere in the surrounding area. They also find physical evidence—indentations in the soil that appear to be from footprints. These observations alone provide them with sufficient information to conclude that what they found in that section of the meadow was the result of someone's plan, and not some natural phenomenon or random occurrence.

It would be fair to say that the observational evidence is conclusive beyond any reasonable doubt. If the hikers investigated further, they might find additional confirmation. Scientific evidence could show that the plants in the parallel rows were all the same kind of vegetable, and not native to the area. Documentary evidence, in the form of a satellite photo from the area, could show the contrast between the cultivated area and the surrounding natural vegetation. By studying the pattern of the footprints, the hikers could provide demonstrative evidence in the form of a diagram showing the footprints coming from a certain direction. There may be testimonial evidence from others who had previously passed through the area. They might then be able to identify who had left the footprints or when they first appeared. A variety of different types of information like this would provide overwhelming confirmation of what had happened in the meadow.

Each bit of evidence is like one piece of a puzzle. A single piece of a puzzle may not reveal much, but when many pieces come together, a picture revealing the truth becomes clear. The physical, scientific, documentary, demonstrative, and testimonial evidence from the cultivated area within the mountain meadow proves to the hikers that someone had planned and planted a vegetable garden.

As this example shows, different types of direct or indirect evidence, compiled from a number of different sources, can often show that a deliberately planned and intentional act had taken place at an earlier time. As you will see, evidence collected in this way reveals the true nature and origin of the universe and leads to some fascinating conclusions. As each piece of evidence corroborates the ones that came before, the picture becomes clear.

Beware of Bias

To arrive at the best possible conclusion, a person should take note of any biases, personal or otherwise, that may affect the decision-making process. It is not easy to set aside long-held opinions or beliefs that may hinder an objective evaluation of evidence.

Throughout history, people have shown a strong desire to defend existing opinions or beliefs rather than questioning them or exploring new possibilities. The ancient Greek historian Thucydides (c.455-c.400 BC) wrote, "For it is a habit of mankind to entrust to careless hope what they long for and to use sovereign reason to thrust aside what they do not fancy."

Beware also of confirmation bias, which is the tendency to cherry-pick evidence and/or accept arguments consistent with a current opinion or belief, while neglecting or rejecting things that do not seem to fit with that belief or opinion. This tendency is very common. Two people with opposing opinions can see the same evidence and feel that it confirms their point of view. Our tendency to take the easiest and quickest path to a decision often compounds the problem of confirmation bias.

Expert witnesses, with knowledge in a field relevant to the case, are often called to testify in court. Expert witness testimony is sometimes given by two different experts, trained in the same area of expertise, each supporting an opposite interpretation of the evidence. Like everyone else, expert witnesses may be biased.

CHAPTER TWO

PHYSICAL LAWS AND PROPERTIES

"The laws of science, as we know them at present, contain many fundamental numbers, like the size of the electric charge of the electron and the ratio of the masses of the proton and the electron …. The remarkable fact is that the values of these numbers seem to have been finely adjusted to make possible the development of life."
Stephen Hawking, physicist and cosmologist

Experiments and Observations

Throughout most of history, human beings only had the five physical senses—touch, sight, sound, smell, and taste—to observe, measure, and study the world and universe around them. New scientific tools, such as electron microscopes, sound analysis equipment, and powerful telescopes have greatly enhanced our physical senses, especially those of vision and hearing.

Previously unknown matter and energy, such as subatomic particles (e.g. electrons, neutrons, and protons) and non-visible light energy (e.g. ultraviolet light and infrared light) can now be studied. Recently developed equipment has made it possible to analyze light and sound energy in great detail.

What was previously undetectable has been exposed in a form that our limited senses of sight and sound can use. Large amounts of scientific data are now available, and we can study previously unknown aspects of reality.

Over the last few hundred years, and particularly during the last few decades, scientific knowledge in the fields of physics, astronomy, chemistry, and biology has grown tremendously. Recent discoveries provide compelling evidence that there is something special about the way that the physical laws and properties of the universe work together.

As many popular television shows and movies have demonstrated, evidence such as a fingerprint can enable a judge or jury to determine who was present and what took place at an earlier time. Another valuable piece of evidence sometimes found is a molecule that contains a unique genetic code for each human being. This fascinating genetic material —deoxyribonucleic acid or DNA—is described in Chapter Four.

Data from physics, astronomy, chemistry, and biology reveals fundamental truths about the universe, despite none of us having been there at the beginning to witness events as they happened. With sufficient quality and quantity of evidence, there is not always a need for eyewitness evidence.

There are two primary types of scientific evidence —experimental and observational. An acceptable theory of reality must be supported by experimental and/or observational evidence. It must be consistent with all available relevant evidence. To ensure a reliable result, evidence must not be discarded or disregarded unless it is shown to be either invalid or not relevant.

Until the last few decades, studies of the properties and makeup of the universe had primarily made use of experimental evidence from physics and chemistry, obtained from repeatable experiments done under controlled conditions.

For example, a simple repeatable chemistry experiment can test the theory that cinnamon oil will kill bacteria. Cinnamon chewing gum, which contains cinnamon oil, can be used to test the theory. Test subjects swab their mouths and place the swab on a Petri dish. This is the control used to measure the number of bacteria in their mouth before they use the gum. After chewing the gum for ten minutes, they swab their mouths again and place the swab on a new Petri dish. After incubating for twenty-four hours, the bacteria are counted using a microscope to determine the effect of chewing the gum. For further confirmation, test subjects can repeat the test with chewing gum containing no cinnamon oil. Several recent studies have shown that, at concentrations of even 10 percent or less, cinnamon oil is effective against Staphylococcus, E. coli, and several other strains of bacteria.

A well-known repeatable physics experiment is the famous Leaning Tower of Pisa experiment, purported to have been performed by the famous Italian scientist Galileo Galilei (15 February 1564 – 8 January 1642) in the late 16th century. This experiment shows that two spheres of different weights, simultaneously dropped from a tower, if not appreciably affected by air resistance, will arrive at the base of the tower at the same time, proving that their time of descent is independent of their weight.

Prior to the 16th century, people had assumed that heavier objects would fall faster than lighter objects. This may be because common lightweight objects are significantly affected by air resistance. The well-known idiom "light as a feather" leads to what seems like a common-sense conclusion that lightweight objects fall slower than heavier objects.

The second common type of scientific evidence is observational evidence. There has been a recent explosion in the quantity and quality of observational data pertaining to the universe. This is primarily due to advancements in astrophysical instruments for observation. With the deployment into Earth's orbit of equipment such as the Hubble Space

Telescope and the Chandra X-ray Observatory, much about the universe has come into focus (literally as well as figuratively).

In the early 17th century, Galileo Galilei constructed a telescope powerful enough to show that the band of light in the sky, now known as the Milky Way, is composed of individual stars. Up until the 20th century, most scientists believed that the entire universe consisted of the Milky Way galaxy. In the 1920s, observations by the American astronomer Edwin Hubble (20 November 1889 – 28 September 1953) proved the existence of galaxies outside our own Milky Way. The largest telescope in the world in 1920 was at the Mount Wilson Observatory in California where Hubble worked. Hubble's observations there showed that objects previously thought to be clouds of dust and gas were entire galaxies separate from the Milky Way galaxy.

The Composition of The Universe

Experimental and observational data has shown that the universe consists of matter along with various forms of energy. The material that makes up any physical object or substance is known as matter. All physical objects or substances consist of combinations of the building blocks of matter known as atoms. Any group of two or more atoms bound together is known as a molecule. Atoms, like the familiar Lego(TM) building blocks, can be joined together to produce complex structures substantially larger than themselves. These structures can be broken down into their component parts, but are not of much use unless combined. Energy in the universe includes electromagnetic energy (the most familiar type being visible light), thermal (heat) energy, and the mechanical energy of moving objects. The physical laws and the properties of this energy and matter in the universe are such that abundant galaxies, stars, planets, and various forms of life not only exist, but persist over time.

One obvious property of the universe is that, due to the nature of gravity, everything exists in a three-dimensional space. In their book, *The Grand Design,* Stephen Hawking and Leonard Mlodinow explain why. "According to the laws of gravity, it is only in three dimensions that stable elliptical orbits are possible. In any but three dimensions even a small disturbance, such as that produced by the pull of the other planets, would send a planet off its circular orbit, and cause it to spiral either into or away from the Sun [2]." For the earth and other planets to even exist, the universe must have three—and only three—dimensions.

Four Fundamental Interactions

One remarkable aspect of the universe is the precise way that matter and energy interact to produce stable elements, compounds, and much larger objects such as stars and planets. The stability of matter is the reason that the universe is not an empty, lifeless void. There are only four known ways that matter and energy in the universe interact. These four fundamental forces, or interactions, must be precisely balanced to maintain the stability of matter in the universe:

- The Force of Gravity
- The Electromagnetic Force
- The Strong Nuclear Force
- The Weak Nuclear Force

The ***force of gravity*** is the most familiar example of a fundamental force. Through the effects of gravity, every object attracts every other object. The magnitude of the gravitational force between two objects depends on the mass of each object and their proximity to each other. Mass is a measure of the amount of material that an object contains.

For example, when a baseball sails through the air after being hit by a bat, the effect of gravity is seen as the baseball falls back to Earth. The force, in this case, exists between one relatively small object (the baseball) and a considerably larger object (Earth). Regardless of the size of any two objects, if the mass of one of the two objects is doubled, the force of attraction will be doubled. The magnitude of the force also becomes much stronger when the objects come close to each other. The force of gravity becomes very substantial for objects with large masses, such as a star or planet.

In a 1687 publication called "Mathematical Principles of Natural Philosophy," English mathematician and physicist Isaac Newton (25 December 1642 – 20 March 1726) described a formula for calculating the force of gravity. Newton related the force of attraction between two objects to their masses and distance apart. As with other fundamental physical laws, Newton's equation included a physical constant, the universal gravitational constant, or G, that can be measured through observation or experimentation. The size of this gravitational constant is such that the force of gravity is not too large or too small, thus ensuring the stability of planets in orbit within our Solar System.

The effect of gravity is a significant factor in the interaction of large objects such as stars and planets. Fortunately, the force of gravity is much weaker than the other fundamental forces, for as astrophysicist Martin Rees explains in *Just Six Numbers: The Deep Forces that Shape the Universe,* "it is only because it is weak compared with the other forces that large and long-lived structures can exist [3]."

The ***electromagnetic force*** generates a force between any particles that each have an electric charge. It can be either attractive or repulsive. It gives rise to magnetism (the force that acts between moving charged particles). This force holds atoms together. Atoms consist of a combination of protons, neutrons, and electrons. Electrons are subatomic particles with a negative electric charge; protons are subatomic particles with a positive electric charge. Protons have a mass more than one thousand times larger than electrons, yet a proton is so small that the dot at the end of this sentence holds more protons than the number of seconds contained in half a million years. Neutrons are subatomic particles similar in size to protons, but with no net electric charge.

Material in the physical world is composed of chemical elements. A full list of these chemical elements is shown in the periodic table of elements found in every high school chemistry classroom. Their structure and behavior are dependent on the properties of the electromagnetic force. An atom of each chemical element can be visualized as being composed of protons and neutrons packed in the central nucleus of the atom, along with one or more electrons separated from the nucleus. Each chemical element can be identified by the number of protons in its nucleus. Figure 1 illustrates the basic structure of a lithium atom which has three protons in the nucleus.

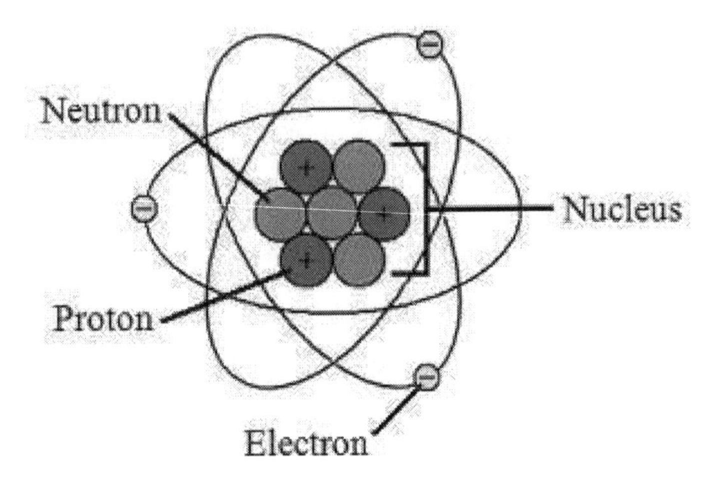

Figure 1. The basic structure of an atom

Due to electromagnetic force, two positively charged particles (e.g. protons) will repel one another. Similarly, two negatively charged particles (e.g. electrons) will repel one another. However, a positively charged particle will attract a negatively charged particle. The force from these electric charges is just right to make possible the stability of the chemical elements.

Fewer than ninety naturally occurring stable chemical elements exist. Unstable elements occur only in small amounts that quickly decay into other elements along with by-products such as protons, neutrons, and electromagnetic energy. An atom of each chemical element shown in the periodic table contains a particular number of protons in the nucleus. However, the number of neutrons in the nucleus may vary. These different varieties of an element are known as isotopes. For example, although all carbon atoms have six protons, the carbon-12 isotope has six neutrons, whereas the carbon-14 isotope has two additional neutrons in the nucleus. As a result, the carbon-14 atom is heavier than the carbon-12 atom. Isotopes are important because they are involved in the nuclear fusion reactions, as described later in this chapter, that produce energy from the sun.

Different from these chemical elements, but no less important, is electromagnetic energy in the form of visible and non-visible light. As with atoms, the characteristics of electromagnetic energy depend fundamentally on the qualities of the electromagnetic force. Types of electromagnetic energy include ultraviolet radiation, visible light, microwaves, radio waves, infrared radiation, gamma rays, and X-rays. Electromagnetic energy is present everywhere and influences every aspect of our lives. We experience the effect of electromagnetism when we are suddenly awakened during a lightning storm, or when we feel the tingling sensation from static electricity.

Magnetism has been observed over many centuries in the form of natural magnets made of the mineral magnetite. A naturally magnetized piece of magnetite, when fashioned into the shape of a pointer and placed on water, could be used as a compass to indicate direction. This naturally occurring magnet was called lodestone, or "leading stone."

It is only in the last 150 years that magnetism and electricity have noticeably transformed our lives. In the 19th century, James Clerk Maxwell (13 June 1831 – 5 November 1879) discovered the connection between electricity and magnetism.

Maxwell, a Scottish scientist working in the field of mathematical physics, studied the effects of electricity, magnetism, and light energy. With the publication of "A Dynamical Theory of the Electromagnetic Field" in 1865, he explained how electric and magnetic energy travel through space as electromagnetic waves, moving at the speed of light [4].

German-born theoretical physicist Albert Einstein (14 March 1879 – 18 April 1955) demonstrated a further astonishing property of electromagnetic energy when he proved the equivalence of energy and matter. This equivalence is expressed by Einstein's famous equation $E=mc^2$ that relates energy "E" to mass "m" and the speed of light in a vacuum "c." This showed for the first time that matter can be thought of as a form of "packaged energy" in a somewhat stable form. The makeup of an "energy bundle," such as an individual atom, will determine the degree of its stability and how the atom will interact with other nearby matter and energy. Electromagnetic force makes this possible.

The ***strong nuclear force*** is much less familiar to people, but also essential to the stability of the chemical elements. It makes it possible for particles in the nucleus of an atom to bind together. It operates over very small distances and at scales much smaller than our senses can detect. It establishes the rules for the behavior of the fundamental building blocks of matter.

The most abundant element in the universe is hydrogen, estimated to make up three-quarters of all elemental matter. This simplest of all elements is composed of a single proton and a single electron. All other elements have more than one proton, and require the presence of the strong nuclear force to permit multiple protons to remain bound close together within the nucleus.

Since protons have a positive electric charge, they will normally repel each other when close together. However, at *extremely* close range, at a distance billions of times smaller than the thickness of a human hair, they will attract each other, due to the strong nuclear force. Without this force, there could not be any elements heavier than simple hydrogen, because two or more protons could not remain together in a nucleus.

The ***weak nuclear force*** is involved in aspects of nuclear decay (the process whereby a nucleus changes by emitting either electromagnetic energy and/or subatomic particles). It makes the nuclear fusion reactions in the sun possible. In many stars including the sun, the temperature is extreme enough so that the protons and electrons no longer form elements such as hydrogen. Instead, the protons and electrons are freed from each other in a state known as plasma. These freely moving particles combine in various reactions that release electromagnetic energy, such as light.

Much of the sun's light energy is produced through proton–proton fusion reactions. As mentioned previously, two positively charged protons will normally repel each other when in close proximity. Proton–proton fusion occurs in stars if the temperature (i.e. kinetic energy, or energy of motion of the protons) is high enough to overcome their mutual repulsion. The weak force, or interaction, permits one of the protons to lose its positive charge by releasing energy in the form of a positively charged particle that quickly combines with one of the many negatively charged free electrons within the sun. This results in a proton and neutron fusing together as one particle, along with the release of electromagnetic energy in the form of light. This new particle will fuse with another proton, resulting in yet another new particle along with the release of still more electromagnetic energy.

Additional fusion reactions create still more particles and energy. Over time, the process converts the original hydrogen nuclei into other heavier nuclei, along with large amounts of electromagnetic energy traveling outward in all directions. This energy continuously radiates out from the sun and fuels life on Earth. All of this is only possible because of the precise nature of the four fundamental forces. Fun fact: The sun fuses 620 million metric tons of hydrogen in its core each second.

In their book *The Grand Design*, Stephen Hawking and Leonard Mlodinow state: "By examining the model universes we generate when the theories of physics are altered in certain ways, one can study the effect of changes to physical law in a methodical manner. Such calculations show that a change of as little as 0.5% in the strength of the Strong Nuclear Force, or 4% in the electric force, would destroy either nearly all carbon or all oxygen in every star, and hence the possibility of life as we know it. Also, most of the fundamental constants appearing in our theories appear fine-tuned in the sense that if they were altered by only modest amounts, the universe would be qualitatively different, and in many cases unsuitable for the development of life. For example; if protons were 0.2% heavier, they would decay into neutrons, destabilizing atoms [2]."

The Expansion of the Universe

Albert Einstein's general theory of relativity revealed that gravity occurring because of the presence of matter and energy will distort space and time. When Einstein developed this theory of gravity while working on the general theory of relativity, his equations showed that the universe should be either expanding or contracting. He and other scientists believed that the universe should instead be static, so he modified his original equations to allow for a static universe.

Several years later, however, Edwin Hubble showed that the universe was indeed expanding. This interesting discovery is explained in more detail in Chapter Three. Measurements have shown that the rate of expansion of the universe is great enough so that matter in the universe will not collapse together from the effects of gravitational attraction. Fortunately for life on Earth, this property of the universe prevents everything, including the Milky Way galaxy, from a catastrophic collapse.

Impossible Coincidences

The physical properties of the universe, along with many fundamental physical constants, such as the universal gravitational constant from Newton's equation of gravity and the speed of light in a vacuum, have the precise values necessary for the universe to be stable and long-lasting. Somehow, the physical laws and properties are favorable for both the existence of fundamental elements and for the combination of these elements into various compounds through chemical bonds. Somehow, the universe also has precisely the right qualities to permit galaxies, stars, and planets to exist in great numbers and to persist over time.

Amazingly, the universe is not only "compound-friendly" in this way, but "bio-friendly" as well. The properties that enable chemical elements and compounds to exist would not necessarily lead to an environment that was favorable to life as well. In fact, for the most part, the properties of the universe are overwhelmingly hostile to life. But we do know that life thrives on at least one planet—our own planet Earth. This bio-friendly nature of the universe is most surprising because it has no natural explanation. In fact, except for a number of factors that work together to benefit and favor life in the universe, there would be a universe full of chemistry, but empty of life.

The universe not only has just the right properties to permit abundant galaxies, stars, and planets to persist over time, but it also has precisely what is needed for life to thrive on at least one planet. What's more, the conditions in some areas of the universe are suitable not only for simple forms of life, but for complex life forms as well.

Consider the seven notable aspects of the universe below:

- The force of gravity is exactly what is needed to ensure the stability of galaxies, stars, and planets within star systems.

- The rate of expansion of the universe is great enough that the force of gravity does not result in the collapse of all matter in the universe.

- The strong nuclear force is exactly what is needed for nuclear fusion to result in a stable nucleus within an atom.

- The weak nuclear force is exactly what is needed to allow successful radioactive decay of subatomic particles and to permit the ongoing hydrogen fusion process in the sun and other stars.

- The magnitude of the electromagnetic force is exactly what is needed to make possible the stability of the chemical elements that are the building blocks for all the visible matter found in the universe.

- The number of spatial dimensions in physical space is exactly what is needed to allow stable orbits of planets to exist in the solar system.

- The masses of the proton, neutron, and electron are exactly what are needed to form stable chemical elements and the isotopes necessary for energy production from nuclear fusion in stars.

An inquiring skeptic requires an explanation for why the universe has these properties that result in more than just the simplest possible form of matter, or just simply light energy, or even nothing at all.

The Requirements for Life

Life can simply be defined as something that has the capacity for metabolism, growth, reaction to stimuli, and reproduction. For life to exist and thrive, certain environmental conditions must be met. The National Aeronautics and Space Administration (NASA) has published a document called the "NASA Astrobiology Roadmap." They define primary habitability criteria for a planet as: "extended regions of liquid water, conditions favorable for the assembly of complex organic molecules, and energy sources to sustain metabolism [5]."

The one chemical compound mentioned by NASA as a necessary ingredient is liquid water. A water molecule has a relatively simple structure composed of two atoms of hydrogen bound together with one atom of oxygen, but it has some unusual properties. Unlike other compounds, it remains liquid over an unusually wide temperature range. The U.S. Geological Survey (USGS) Water Science School states that, "Water is unique in that it is the only natural substance that is found in all three physical states—liquid, solid, and gas—at the temperatures normally found on Earth [6]."

Water is liquid over a temperature range of about 100 degrees Celsius (from 0 to 100 degrees). For comparison, hydrogen itself is liquid over a range of about 6 degrees Celsius (from -259 to -253 degrees) and oxygen is liquid over a range of about 36 degrees Celsius (from -219 to -183 degrees). Another common compound is carbon dioxide. It is liquid over a temperature range of about 21 degrees Celsius (from -78 to -57 degrees).

The USGS Water Science School publication points out another special property of water. "Water is called the universal solvent because it dissolves more substances than any other liquid. This is important to every living thing on Earth. It means that wherever water goes, either through the air, the ground, or through our bodies, it takes along valuable chemicals, minerals, and nutrients."

Perhaps the most amazing property of water is its behavior near its freezing point. Most substances are more dense as a solid than a liquid. Water is unusual in that its solid form, ice, is less dense than its liquid form. Liquid water shrinks as it cools until it reaches 4 degrees Celsius, then it begins expanding. Because of this property, it is one of the few materials that floats when in its solid form. Without this property, freezing rain, frozen snow, and ice would sink and accumulate in lakes and in the ocean depths. The oceans would freeze from the bottom up.

For life to be sustained, there must be a persistent and sufficiently large energy source. The universe provides an energy supply for star systems— such as our own Solar System—through radiant energy from the stars within those systems. This energy is supplied in the form of electromagnetic energy through the process of nuclear fusion.

The following elements are necessary for life (simple or complex) to exist on a planet:

- An atmosphere, to prevent extreme variation in temperature between night and day and to provide a shield of protection from incoming genetically destructive ultraviolet radiation. Light energy from the sun consists of visible light, but also low energy infrared and high energy ultraviolet radiation. The low energy infrared radiation is not harmful to life on a planet, but exposure to the high energy ultraviolet radiation will suppress the immune system, cause premature aging of the skin, damage the eyes, and even cause skin cancer. The predominant gases in Earth's atmosphere are nitrogen, 78 percent, and oxygen, 21 percent. An important aspect of the upper atmosphere is the presence of ozone, also known as trioxygen, a molecule formed by the combination of three oxygen atoms. It is much less stable than the common form of oxygen composed of two oxygen atoms. The oxygen-rich atmosphere provides a layer of ozone (also called the ozone shield) as protection from dangerous ultraviolet radiation. Some ultraviolet radiation is absorbed by the nitrogen and the common form of oxygen in the atmosphere. However, some of the most dangerous ultraviolet radiation can only be adequately shielded by the ozone. In the ozone layer, the number of oxygen molecules of the common form is over one hundred thousand times the number of ozone molecules. Fortunately, there are enough of these scarcer ozone molecules to do the job of absorbing most of the dangerous ultraviolet radiation emanating from the sun.

- Six elements are necessary for life. CHNOPS stands for the chemical abbreviations of carbon, hydrogen, nitrogen, oxygen, phosphorus, and sulfur. Combinations of these six elements make up most of the biological molecules on Earth. They make up almost 99 percent of the mass in the human body and are necessary for the existence of organic compounds. If any one of them were not present in the universe, or not stable over time, life as we know it could not exist.

- There must be a source of adequate energy, such as the radiant electromagnetic energy supplied by a star. The star must also have sufficient gravity to hold the planet in a stable orbit.

- A large iron core supplies the planet with a magnetic field to protect it from radiation that would be deadly to life and would eventually strip away the atmosphere. A star such as the sun emanates a constant stream of dangerous high energy electrically charged particles known in our Solar System as the solar wind. The movement of molten iron deep within a planet such as Earth creates electric currents that generate a protective magnetic field to shield against these particles. Through a complex process, the magnetic field breaks down and redirects the particles, so that only a small amount of the radiation penetrates the shield.

- The rotation period of the planet must be great enough to create the necessary protective magnetic field. A moving electric charge will produce a magnetic field. To demonstrate this effect, we can make a simple electromagnet by wrapping a wire around an iron bar, then passing an alternating electric current through the wire. The motion of electrons in the wire will produce a magnetic field. An electromagnet like this will attract certain metals such as iron and nickel, but not other metals such as copper. A planet's magnetic field is created by swirling motions of liquid-conducting materials, such as iron, in the planet's interior. If the interior of a planet is solid, or the speed of rotation is too slow, the magnetic field will be too weak to provide protection for the planet.

- A sufficient supply of liquid water on its surface is necessary.

- The planetary orbit must be in what scientists have defined as a "habitable zone" within a star system. This requires that the planet be close enough to the star so that water will exist in a liquid form, but not so close as to destroy any protective atmosphere and boil away all the elements and compounds necessary for life.

- The planet's orbital motion needs to be nearly circular so it is entirely within the habitable zone during its journey around the star.

Some of these conditions, such as an energy source and protection from deadly cosmic radiation, are absolutely essential. The absence of one or more of the others would produce an environment unlikely to permit the existence of complex organic molecules. The fact that they are ***all*** permitted confirms the "bio-friendly" nature of portions of the universe.

Life would be impossible if the universe did not permit these conditions to exist.

It cannot be concluded that Earth is unique in the universe, for there are many star systems similar to our Solar System that could support Earth-like planets. However, considering the many aspects of the universe that are unfavorable for life, it is remarkable that the conditions necessary to sustain life exist anywhere in the universe. The special bio-friendly nature of the universe is only possible because of the precise nature of the physical laws and properties of the universe.

The preciseness of the characteristics of the universe is comparable in many respects to the isolated mountain meadow discovery described previously. Chapter Three explores this further. From the form, structure, and setting, the only reasonable explanation is that the hikers found an intentionally constructed garden plot. There is no alternative explanation for the neat rows, or why so few of the native plants from the surrounding area are in the garden. They cannot conclude that they have simply come across a rare coincidence. The evidence clearly shows that someone had planned and planted a garden in that plot.

CHAPTER THREE

EVIDENCE CHALLENGES THE STATUS QUO

"A wise man proportions his belief to the evidence."
David Hume, philosopher

Misled by the Experts

We know that the form and substance of the universe permit life, even intelligent life, to thrive and prosper. From the evidence, we must conclude that either there is a deliberate plan to the workings of the universe, or that matter and energy spontaneously came into existence with the necessary properties for the entire universe to be stable over long periods of time, and with the right properties for life to thrive and prosper. The simple, straightforward explanation, as in the case of the isolated mountain meadow, is that it has been planned that way.

Throughout human history, people seeking answers to questions about the nature and origin of the world and universe around them have had only one source to rely on: the claims and beliefs of those whom they admired or revered within society. They would get some answers, but often with little or no support from real evidence. Without any real evidence, one person's idea is as good as another.

Incorrect conclusions are often based on admiration—or disdain—for someone who holds a particular belief or behaves in a certain way, such as appearing knowledgeable or confident. Preconceived ideas, former well-established opinions, or long-held beliefs of experts of the day have often misled people.

In any area of study, there are "experts" who act self-assured, but their statements may be far from the truth. They can appear quite confident in their educated guesses or "informed opinions." They may arrive at conclusions before they consider all the available information. Experts usually have a narrow field of knowledge and have not considered observations or data from other areas. They may not even be aware of, or had time to investigate, recently discovered results from experiments. Do not mistake confidence for competence.

Even nowadays, simply accepting what the scholarly experts have concluded often leads to an incorrect result. They may have a personal bias that favors a certain opinion or belief, with little interest in investigating new information that may be contrary to their current opinion or belief. This is, of course, not an uncommon trait among the general population. Throughout history, experts of the day have brushed aside newly discovered scientific evidence. There is a natural desire to avoid any threat to long-held ideas, and thereby to preserve the current status quo.

People frequently make decisions based on what makes the most sense to them, or by simply accepting what the "learned experts" have concluded. Reaching a conclusion this way, rather than taking time to gather and examine evidence, is relatively fast and easy but generally unwise. What makes sense to one person can be quite different from what makes sense to another, due to differing past experiences and personal views. Failure to evaluate evidence, or simply accepting the opinions and beliefs of others, can lead to wrong conclusions, and often unwise decisions.

Learning from History

Greek philosopher and scientist Aristotle lived over 2000 years ago. Then, as now, everyone saw the sun rising in the east and setting in the west. It seems obvious from common sense that the sun must be moving above a stationary Earth. Aristotle was one of the first to formally propose and describe in detail a geocentric universe where Earth exists at the center, surrounded by concentric celestial spheres of planets and stars. Few people questioned this obvious conclusion. This example shows once again, as in the case of the Leaning Tower of Pisa experiment, that "common sense" does not always lead to the right conclusion.

A contemporary of Aristotle, Aristarchus of Samos, was an ancient Greek astronomer and mathematician. He proposed a model of the known universe with the sun at the center and the earth revolving around it. However, he was unable to provide any supporting data at the time. As a result, not many people were convinced that he was right.

His idea was not pursued further until more than 1700 years later. The mathematician and astronomer Nicholas Copernicus (19 February 1473 – 24 May 1543) had developed a model of the universe that placed the sun, rather than the earth, at the center. Although the telescope would not be invented until nearly seventy years later, Copernicus managed to use his observations of the sky to show that the earth did indeed appear to revolve around the sun. Copernicus published a book shortly before his death in 1543. He described his model and explained his theory that the earth moved across the heavens, orbiting the sun in a similar fashion to the other planets. Unfortunately, despite the supporting evidence, experts at the time rejected his model of the universe.

In the following century, observations made by Galileo Galilei provided additional support for the Copernican model. Galileo Galilei was a multi-talented scientist known for his work as an astronomer, physicist, engineer, philosopher, and mathematician. In 1609, Galileo built his version of a telescope with which he observed that the planet Venus had phases similar to the well-known phases of Earth's moon. Galileo observed that the phases of Venus were consistent with the model proposed by Copernicus. Once again, the experts at the time rejected the new evidence. The accepted scientific and philosophical ideas about the universe, based on the writings of Aristotle and other ancient Greeks, remained firmly entrenched. The widespread acceptance of Galileo's ideas took many years.

Another example of experts of the day brushing aside newly discovered scientific evidence occurred in the 17th century. William Harvey (1 April 1578 – 3 June 1657) was an English physician who studied, in great detail, blood circulation to the heart, brain, and body. In 1628, he published his findings describing the structure of the heart and arteries. He suggested that blood passed through the heart, not the liver as previously believed. Harvey's findings were ridiculed, and many doctors scoffed at the idea. His own medical practice declined because of the barrage of criticism from his fellow physicians. In his later years, ostracized by the scientific world, Harvey became a recluse. Harvey's work on circulation became accepted after many years had passed.

A more recent situation occurred in the 19th century, when many lives were lost because of the rejection of evidence presented by Hungarian physician Ignaz Semmelweis (1 July 1818 – 13 August 1865). While working in Vienna General Hospital's First Obstetrical Clinic, he observed that doctors' wards had three times the mortality of midwife wards. In 1847, he proposed the practice of washing hands with chlorine lime solutions. Even though washing hands could save lives, the medical community rejected his ideas. Semmelweis's observations conflicted with the accepted scientific and medical opinions of the time. Before the 19th century, doctors were not aware that bacteria spread through a lack of sanitation. Semmelweis's suggestion earned acceptance only years after his death, when the French biologist and chemist Louis Pasteur (27 December 1822 – 28 September 1895) confirmed the germ theory, and British surgeon Joseph Lister (5 April 1827 – 10 February 1912) began successfully using hygienic methods in his surgery.

The experience of William Coley (12 January 1862 – 16 April 1936) provides an example from more recent history. Dr. Coley worked at New York Cancer Hospital as a cancer researcher and bone surgeon. Coley noticed that some patients began recovering from cancer after being infected with the same bacteria that causes strep throat. Standard cancer treatment involved surgically removing the cancerous tissues; no cancer drugs or radiation therapy were yet available. Coley theorized that post-surgical infections helped defeat cancer by mobilizing the immune system. Most of his scientific peers rejected the idea of deliberately introducing potentially deadly bacteria into patients, writing it off as crazy and dangerous. He died in 1936, without ever knowing that his work would lead to the birth of modern immunotherapy.

As recently as the 1980s, Australian physician Barry Marshall's research showed once again that preconceived ideas and long-held beliefs of scientific experts can impede progress towards discovering the true nature of things. Barry Marshall (born 30 September 1951) discovered that bacteria appeared to cause ulcers and stomach cancer. In the 1980s, doctors believed that stress caused ulcers. When Marshall suggested that antibiotics could be used to treat ulcers, his fellow physicians ridiculed the idea. They insisted that bacteria could not live in the acidic environment of the stomach.

To finally prove his theory, Marshall mixed the ulcer-causing bacteria into a broth and drank it himself. He became ill, as he had predicted, and showed that the bacteria can live and thrive in the stomach. More importantly, he confirmed that antibiotics can treat the bacteria that causes ulcers. Standard care for ulcers now involves antibiotics to kill off the bacterial infection. Thanks to his research, stomach cancer also is no longer one of the most common forms of malignancy. In 2005, Barry Marshall and his fellow researcher J. Robin Warren received the Nobel Prize for their discovery that stomach ulcers are caused by bacteria.

These examples illustrate the importance of taking a serious look at newly discovered evidence with an open mind and clear eyes.

Revelations from the Physical Laws and Properties of the Universe

Scientists, including cosmologists who study the large-scale properties of the universe, have, for decades, analyzed huge volumes of experimental and observational data. They have not been able to find evidence to explain the special qualities of the physical laws and properties of the universe. Somehow, there must be an explanation for the fact that the building blocks of matter have such precise properties, and that the laws of nature work in just the right way, to sustain everything.

The scientific community has spent an enormous amount of time and money looking for an explanation that does not involve some sort of plan. It would be simple for scientists if the properties of the universe were the result of some natural physical necessity, but nothing has been found to support that idea. No evidence exists that the physical constants, such as the universal gravitational constant in Newton's equation of gravity, and other properties of the universe, are required through some fundamental natural law to be what we measure them to be. Even if the characteristics of the universe were somehow determined by some unexplained means, there is no reason to expect that the universe would also turn out to be stable and long-lasting, and to have the properties required to support life.

A philosophical idea called the Anthropic Principle has been proposed by some in an attempt to explain why the physical laws and properties must be as we find them. The most common form of the Anthropic Principle states that, because we humans are here, it follows that the physical laws and properties of the universe must be favorable for our existence. It is true that if the physical laws and properties were not favorable, we could not be here. However, the principle only confirms that the physical laws and properties are sufficient for life to survive and thrive in the universe. It cannot be concluded that they are as we find them ***because of*** our presence here in the universe. The more correct statement is, "Because we are here, it follows that the existing physical laws and properties of the universe ***are favorable*** for our existence," not that they ***must*** be favorable for our existence.

The analogy of an automobile race will help to clarify this point. Because a certain vehicle won the race, it follows that the mechanical properties of the vehicle prior to the race must have been favorable for the win. It does not follow that the mechanical properties ***are the result of*** the car winning the race. In fact, the mechanical properties are the result of a great deal of planning and engineering prior to the race.

The Anthropic principle is like answering the question, "Why is there oxygen in Earth's atmosphere?" by saying that it is because we can live on Earth and people need oxygen to live. This statement, though true, does not answer the original question, "Why is there oxygen in Earth's atmosphere?" It instead answers the simpler question, "How do we know that there must be oxygen in the atmosphere?" Giving the answer to a similar question may produce a feeling of satisfaction, but the original question remains unanswered.

Astrophysicists Bernard Carr and Martin Rees note, "even if all apparently anthropic coincidences could be explained (in terms of some grand unified theory), it would still be remarkable that the relationships dictated by physical theory happened also to be those propitious for life [7]."

It is impossible to comprehend the huge balancing act required for the properties of the universe to be what we observe. There is no experimental or observational evidence to support an alternative to a deliberately planned universe. Cosmologists have come to the realization that there is only one way that the universe could possibly be unplanned. They have devised a model of reality that requires a nearly infinite number of undetectable universes that have somehow appeared spontaneously, each with different physical laws and properties. In this model, our universe is one of them with the correct combination of laws and properties that results in what we observe. Cosmologists call this the multi-universe, or multiverse. As impossible as it seems, and despite having no evidence to support the idea, many scientists now hold to the speculative notion that our universe is one of many that have somehow appeared spontaneously from nothing.

This idea suggests that the universe came into existence from nothing, as the result of random chance, through a process that somehow generated a nearly infinite number of very different universes. Among this multitude of universes, a robust, long-lasting, and stable universe appeared, with just the right conditions to allow the sustainable development and expansion over time of everything that we observe. This, of course, requires an incredibly precise balance of forces, matter, and energy. This stable universe also appeared with the right conditions to allow life—including intelligent life—to exist and thrive. It also somehow produced living things from inanimate matter through some unknown means. In other words, it was not only life-permitting, but had just the right characteristics to be life-producing as well.

The multiverse idea requires that a physical reality consisting of other universes exists outside the realm of the known universe. Each different universe has somehow remained isolated, so that observers are not able to detect the presence of the nearly infinite number of other universes. The universes must not interact with each other in any way, so that they remain undetectable even after long periods of time have passed. The multiverse idea presumes the existence of a nearly infinite number of unobserved universes to try to explain the existence of the one observed universe.

All of these universes need widely varying properties, so that at least one of them has precisely the properties that we observe. Because of the huge number of variables in the multiverse idea, the number of different universes must be virtually infinite. Stanford physicists Andrei Linde and Vitaly Vanchurin have calculated the number of universes contained within a multiverse. The result is a number that would be written out by a numeral one followed by ***billions of zeros*** [8]!

This is a number impossible to comprehend. One of the longest white sand beaches in the world is the "Ninety Mile Beach" in Australia. Think of the number of sand grains that one would walk past and over during the journey of ninety miles (151 kilometers) along this beach. This number represents only a small fraction of all the sand grains on the rest of the world's beaches. People have estimated the number of sand grains on all of the world's beaches to be 1,000,000,000,000,000,000. The total number of sand grains on Earth is but a tiny fraction of the number of different universes required by the multiverse idea.

No experimental or observational data exists to support the idea of universes other than our own. The evidence that does exist reveals that an unplanned universe is not possible unless a bizarre, unimaginably complex combination of circumstances were to occur. The multiverse is an idea devised to explain away the overwhelming existing evidence that points to a planned universe.

If the hikers in the mountain meadow had to come up with an explanation for the cultivated garden plot, other than a planned construction, they would find it difficult. How could it be an unplanned natural occurrence? There would have to be a huge number of widely scattered garden plots with various kinds of plants that spontaneously appear in the mountains—each having different shaped plots and configurations. The hikers just happened to come upon a garden plot with the correct shape, layout, and combination of plants in straight parallel rows. But what about the indentations in the soil that appear to be footprints? One hiker might suggest that they were caused by a hail storm that had passed through the area. The hailstones had randomly fallen in a pattern of indentations resembling footprints. The probability of that happening is, of course, ridiculously small.

An important requirement for all scientific theories or propositions is that they be verifiable or falsifiable through experimental and/or observational evidence. The multiverse proposition fails to offer any method to test it. A proposition that makes no observable predictions is not a useful theory. It is no more than someone's speculative notion.

Sir Thomas More (7 February 1478 – 6 July 1535) was an English lawyer, social philosopher, author, and statesman. While imprisoned in the Tower of London in 1534, he wrote "A Dialogue of Comfort against Tribulation," shortly before his execution. The king had ordered More's execution because of More's refusal to swear allegiance to the monarch (King Henry VIII), as Supreme Governor of the Church of England. He wrote, "like a man that in peril of drowning catcheth whatsoever cometh next to hand, and that holdeth he fast, be it never so simple a stick; but then that helpeth him not… for that stick he draweth down under the water with him, and there lie they drowned together." Most people will recognize this as the modern idiom "grasping at straws."

The scientists that want to believe the multiverse exists are grasping at anything in a futile attempt to save their idea from sinking, but can offer no evidence for their speculation. Without evidence it remains a fanciful idea, but something substantive is needed before declaring it a real possibility.

Spontaneous "From Nothing" Idea

Scientists have made great progress in understanding how the universe works. It is not necessary to know where something originally came from—or how it came into existence—to study, analyze, and ultimately understand how it works.

For example, the invention of the airplane originally came from the ideas of men such as brothers Orville and Wilbur Wright. After that, Frank Whittle of the United Kingdom and Hans von Ohain of Germany both worked on the development of the jet engine during the late 1930s. If intelligent beings from another planet were to come upon a modern aircraft, they could study it to understand its characteristics and how various parts of it work. They certainly would never conclude that the aircraft originally came into existence from nothing!

The origin of the universe has the scientific community stumped. Incredibly, cosmologists suggest that something immensely more complex than an airplane, namely the entire universe, came into existence spontaneously— from nothing—along with a nearly infinite number of other universes.

In their book *The Grand Design*, Stephen Hawking and Leonard Mlodinow declare, "As recent advances in cosmology suggest, the laws of gravity and quantum theory allow universes to appear spontaneously from nothing. Spontaneous creation is the reason there is something rather than nothing, why the universe exists, why we exist [2]."

Note that they do not claim that the laws of gravity and quantum theory *require* that universes appear spontaneously from nothing. The statement claims only that scientists have devised a quantum theory that *may allow* for the possibility of universes to come into existence from nothing. Without verification from experimental or observational data, one should not conclude that something exists because a current theory permits it as a possibility.

For example, a biologist might claim that the principles of cell theory allow for the possibility of a fruit that looks like a banana but tastes like an orange. Such a fruit can be imagined, but we have no evidence that it exists in reality, despite a thorough knowledge and exhaustive study of all fruit-bearing plants on Earth. Without verification from experimental or observational data, it cannot be concluded that such a fruit exists.

Mathematics can help us understand and describe the way that the universe works. However, if mathematical equations are developed that suggest that something could be theoretically possible, it does not then follow that it must in fact be true. Mathematics can be misleading if not supported by real-world data. If the equations do not fit with what is observed, one must look for a different explanation.

Spontaneous Creation proponents have started with the presupposition that there can be no intentional planning, and then produced an idea of reality for which there is no observed evidence. Physicists have failed to follow where the data leads. It might be appropriate to label this "Fantasy Physics."

Cosmologists cannot find experimental or observational data to explain what is behind the special qualities of the universe. Instead of accepting the straightforward explanation that the universe is planned to be the way that it is, they attempt to create an explanation to fit with their current point of view.

The fact that many people favor an opinion cannot be used as evidence to support a theory. As discussed previously, there are examples throughout history where most experts of the day mistakenly held on to current opinions or assumptions. Sometimes, it required a whole new generation of younger scientists willing to consider other possibilities before people would accept the evidence that more correctly described reality.

Evidence Prevails

Fortunately, there are examples from history where new evidence forced aside resistance to change. Three of the most notable examples are the belief in "spontaneous generation" in the field of biology, the assumption of a spatially and temporally infinite static universe in cosmology, and the belief in luminiferous ether in physics.

Given the belief by some within the scientific community that the universe appeared spontaneously from nothing, one of the most interesting of all examples from history was the belief in "spontaneous generation."

Spontaneous generation is the idea that life can arise from nonliving matter. The sudden appearance of fish in a new but isolated puddle of water, and the appearance of maggots on meat left out in the open air, seemed to have no other explanation. Aristotle was one of the first to support and describe the idea of spontaneous generation. As impossible as it now seems, for hundreds of years people accepted this idea as a scientific fact. As recently as the 18th century, some scientists continued to support the idea.

In the 19th century, the biologist Louis Pasteur suspected that fermentation was caused by living microorganisms. Most of his contemporaries believed that fermentation was spontaneously generated through a series of chemical reactions. Pasteur vigorously fought against the theory of spontaneous generation. He demonstrated that in sterilized and sealed flasks containing boiled liquid, no living organisms appeared. However, in sterilized open flasks, microorganisms did appear. Additional experiments that he and others performed provided more evidence that finally dispelled the notion that life can arise from nonliving matter.

The scientific community accepted the idea of a spatially and temporally infinite static universe until the 20th century. As mentioned previously, Albert Einstein developed the general theory of relativity that suggested that the universe should be either expanding or collapsing. But, because he and other scientists at the time presupposed that the universe must be a spatially and temporally infinite static universe, he made a change to his theory so that it would work with a static universe.

A few years later, in the 1920s, Edwin Hubble showed that galaxies in the universe are moving away from each other—the more distant the galaxy, the faster it is moving away from us. The expanding universe is like a loaf of raisin bread rising in the oven. As the bread expands, the distance between each raisin in the bread steadily increases. Similarly, galaxies become further separated from each other as the universe expands. The way that Hubble described the expanding universe is known as Hubble's Law.

Hubble's Law revealed that the universe is expanding as predicted by the original formulation of the general theory of relativity. As a result, Einstein acknowledged that his modification of the original general theory of relativity had not been necessary. Though happy to learn that an explanation had been found, Einstein referred to his original adherence to the old idea of the static universe as his "greatest blunder."

Before the last century, most scientists believed that light energy propagated through the empty vacuum of space in an undetectable, invisible medium known as the luminiferous ether. The ether provided the medium through which electromagnetic waves traveled, similar to the way that water provides the medium for ocean waves to travel from one place to another. The scientific evidence up to that time seemed to point to that conclusion.

In 1873, James Clerk Maxwell described the properties of electromagnetic waves in "A Treatise on Electricity and Magnetism." He suggested that these waves were similar to light, and traveled in the same manner as light through the luminiferous ether. Subsequent laboratory experiments proved the existence of such waves, and their speed was indeed found to be the same as the speed of light. We now know that visible light is but one form of electromagnetic wave.

In 1887, physicist Albert Abraham Michelson (19 December

1852 – 9 May 1931) and chemist Edward Williams Morley (29 January 1838 – 24 February 1923) designed a scientific experiment to confirm the presence of the luminiferous ether. Since the theory held that this ether permeated the entire universe, they set up the experiment to measure the speed of light in the direction of the earth's motion as it travels in its orbit around the sun. This speed was then compared to the speed of light at right angles to the earth's motion. To the surprise of the scientific community, the experiment showed no difference in the measured speeds.

If the light had propagated through the ether, the speed relative to the earth should be different when measured in different directions. Most of the scientific community could not initially bring themselves to accept this result. Michelson felt that his experiment was a failure, and for the rest of his life he continued to believe the experiment had failed. The result of his experiment became widely accepted only after Einstein's general theory of relativity revealed that the speed of light has the same value when measured by different observers in uniform motion relative to each other.

In her book *The Hole in the Universe*, K.C. Cole probably said it best: "Indeed, this so-called luminiferous ether gave physics such a black eye that it's often invoked as an example of the way physicists sometimes seduce themselves into believing the inventions of their own fertile minds, even when nature provides ample clues to the contrary [9]."

Fantasy Physics Fails

We are now witnessing another example, as many scientists continue to support the opinion that there is no underlying plan to the universe, despite the mounting evidence to the contrary. Like the disproved spontaneous generation explained previously, they hope to explain away something that they do not understand and that does not agree with their current viewpoint.

The universe is like a machine in which every component must work together perfectly for the machine to work properly. However, it cannot be concluded that, due to the great complexity of the universe, its origin must have required great planning. Complexity itself does not provide real evidence. Many of the physical laws and properties are not all that complex, though they may not be easy to comprehend. Fortunately, we understand enough from the available evidence to reach valid conclusions about the nature and origin of our world and the universe.

Speculative ideas such as “spontaneous creation from nothing” and the “multiverse,” with no supporting experimental or observational evidence, indicate that the correct explanation must be found elsewhere. Almost surprisingly, it is revealed not only by the physical laws and properties already described, but also by additional information found in other areas of study, as detailed in the following chapters.

The universe is not a jumble of matter and energy that randomly popped into existence from nothing. The special qualities of the universe point to only one reasonable conclusion: the universe is the result of an intentional plan. It is the only explanation supported by data and observations. For further valuable insight, let's explore the realm of living things and examine the operation of digital computer systems.

CHAPTER FOUR

REVELATIONS FROM LIFE

"The more I examine the universe and the details of its architecture, the more evidence I find that the universe in some sense must have known we were coming."
Freeman Dyson, physicist and mathematician

Human and Animal Characteristics

It is not only experiments and observations from the physical sciences that provide evidence of intentional planning. We find more evidence when we investigate the unusual and extraordinary aspects of the various forms of life found on planet Earth. Scientists have cataloged over one million species of living things on Earth. Researchers report that more than fifteen thousand new species are discovered each year. Camilo Mora, a marine ecologist at the University of Hawaii at Manoa, and his colleagues at Dalhousie University in Halifax, Canada, have estimated that the total number of species on Earth is well over seven million [10]. The various forms of life on Earth include plants, animals, fungi, and single-celled organisms such as bacteria. This chapter investigates the extraordinary differences between animals and human beings. For clarity, I will use the simple term "animal" to refer to the non-human multicellular eukaryotic organisms on the earth.

An animal's need for survival and propagation determines its behavior in a mostly predictable manner. Animals have physical qualities and instincts well-suited for those needs. They also learn from experiences when it benefits their survival or propagation. Unlike the other living things on Earth, human beings exhibit behavior that extends far beyond the basic needs for survival and propagation. For example, humans are “creators.” Human beings learn and participate in activities that result in personal satisfaction. They enjoy artwork, crafts, reading, and music.

Only humans can develop fields of study in, and practical applications from:

- Arts and Humanities (archaeology, architecture, art, law, history, music, literature, linguistics, performing arts, and philosophy)

- Social Sciences (education, accounting, commerce, mathematics, marketing, agriculture, economics, management science, political science, psychology, and environmental studies)

- Engineering (chemical, civil, structural, electrical, mechanical, aeronautical, and mining)

- Science (biology, physics, chemistry, astronomy, earth science, computer science, life science, and medical science)

Consider some of the extraordinary human abilities and talents absent in ***all*** animals:

- Using language and symbols to communicate and share ideas:
  Humans use conversational language for exchanging ideas, opinions, sentiments, or observations. They use humor and laughter to express mirth, pleasure, or nervousness. They write to communicate thoughts,

ideas, or concepts by using symbols for reading, writing, and mathematics.

- Using clothing:
  Animals show no interest in clothing, no matter how cold or wet they may be.

- Creating and enjoying entertainment:
  Humans participate in the performing arts to convey artistic expression.
  They enjoy visual arts and crafts, such as drawing, painting, architecture, or sculpting. They write fictional and non-fictional stories to relate events to each other, as in a story narrative. They create music and use musical instruments to combine sounds for pleasure or enjoyment.

- Using fire:
  Only humans use fire for lighting, heating, cooking, producing steam, conducting ceremonies, or sending smoke signals.

- Using money or other objects for the exchange of goods and services:
  Animals do not use items or objects found in their environment as a medium of exchange to obtain items from other animals.

- Inventing new things to improve their lives:
  Humans create or produce new things to assist themselves or others with their daily living such as

electric lights, automobiles, airplanes, telephones, digital computers, and the internet.

- Developing instruments to enhance the five senses: Humans build devices such as the microscope, telescope, or sound amplifier. When interacting with each other and their environment, animals rely solely on their inborn physical senses and intuition. The only exceptions are the result of deliberate programming or behavioral training by humans.

- Integrating knowledge from different areas of study to better understand the material world:
  Animals are not able to expand their understanding of the world around them by combining different types of information.

All animal species— more than seven million on this planet—despite widely varying physical and mental characteristics, have one thing in common. They lack ***all*** of these uniquely human abilities. That one type of living being should possess all of them, while all other types possess none, is remarkable. Something extraordinarily special is going on.

There are those who say that humans are arrogant to compare these exceptional qualities of human beings to those of animals. However, these differences do exist and must be acknowledged if we are to progress towards a correct understanding of reality. These many differences between human beings and animals do not indicate that humans are in every way better or superior than animals. In many ways, such as instinct, sense of smell, and eyesight, humans are clearly inferior to many different animals.

Despite having certain special qualities, human beings have no reason to feel proud. In many ways, humans are not at all better or superior to animals. However, it cannot be denied that human beings differ from animals in many ways—some good and some bad.

Certain animal behaviors exhibit similarities to some human abilities. Huge expenditures and years of effort by highly educated people have been devoted to the study of animal behavior. However, no conclusion has been established to support the idea of animal cognition.

Tool Use

Animals use items found in their environment, such as sticks or stones, to throw, jab, reach, pound, drop, block, pry, wipe, rub, or dig when foraging for food. Chimpanzees are known to pick up various items in their surroundings to help them obtain food. With many hours of effort by human handlers, they can even be trained to use gestures or non-verbal signs to indicate what they want, such as an item of food. Surprisingly, despite the advantages of considerably larger brains and an ability to easily grasp objects, even animal species such as primates are no more effective with tool use than some species of birds.

For example, the crows from the Pacific island of New Caledonia, located east of Australia, are on a level with chimpanzees when it comes to finding novel uses for items from their natural environment. Like chimpanzees, they use objects such as twigs or leaves to probe crevices for food, such as insects. However, when researchers put food out of their reach, the crows sometimes used a short stick to retrieve a longer stick that they then used to obtain the food. "It was surprising to find that these creatures performed at the same levels as the best performances by great apes on such a difficult problem," said Russell Gray, of the University of Auckland in New Zealand [11].

Other experiments with Caledonian crows were done by researchers from the Department of Zoology at the University of Oxford in England. The researchers there reported that "seemingly intelligent behavior can be achieved without the involvement of high-level mental faculties, and detailed analyses are necessary before accepting claims for complex cognitive abilities [12]." Scientists consider primates to be a significantly more advanced species than crows, and few would dispute that conclusion. This conundrum is investigated further in Chapter Five.

The bolas spider demonstrates a fascinating example of apparent tool use by an animal that we would consider having no intellect. Bolas spiders are found in America, Africa, and Australasia. They feed mostly on moth species that have scaly wings like those of butterflies. These wings allow them to slip out of most spider webs. An unusual technique enables the bolas spider to effectively catch its prey. The bolas is a throwing weapon used by South American gauchos. Weights on the ends of interconnected cords are specifically designed to capture animals by entangling their legs. In a similar manner, the bolas spider uses a sticky blob of silk on the end of a silk line. The spider swings this at a flying insect, such as a moth that is unfortunate enough to pass nearby. The spider snags the flying insect and reels it in, much like a fish on the end of a fishing line. The spider did not invent this device after some careful study and analysis of the principles of mechanics.[1] This innate behavior is part of the spider's makeup and very effective for ensuring survival in its environment.

Having a larger brain, a complex nervous system, or greater physical strength does not necessarily indicate that one animal species will be more successful at survival than another. The tiny bolas spider is no less successful at survival within its environment than are other animal species within their environments.

Animals simply use a nearby item to accomplish some task that helps them to achieve one of their two purposes in life—survival and propagation. There is a huge difference between an animal that picks up a stick and pokes it in a hole to get at some food and a human who fashions a bow and arrow from sticks and twine for hunting. Similarly, there is a big difference between an animal that picks up a rock and throws it to acquire some food and a human who picks up a rock and uses it to sharpen a stick to use as a spear. “Animal use of tools” should more properly be referred to as “animal use of found items,” or perhaps “animal use of objects.”

---

[1] Mechanics is a branch of physics used to describe and predict the movement of objects subjected to and acted upon by forces.

Mental Abilities and Natural Talents

It is not surprising that animals will learn through experience that some kinds of behavior can result in an improvement in their success with survival or propagation. Chimpanzees appear at times to practice deception to get what they want, but other less capable animal species will also practice deception at times. For example, some mother birds will feign a broken wing to lure a predator away from their nest. Male monkeys and squirrels will deceive other males when competing for a female.

Animals do communicate, but they do not converse. Conversation and the sharing of ideas is uniquely human. Disputes and conflicts among animals certainly do exist, but there are no discussions or debates in the animal kingdom. Animals make certain sounds that may distract predators, signal danger to their neighbors, or call for a mate, but nothing like a conversation exists in the animal kingdom.

A conscious, logical choice or decision requires the use of the concept of time and an understanding of causation. Human beings can use logic and reasoning, rather than instinct, to make choices and answer questions. They can think about what others are thinking, and have the ability to escape from their present point of view. They can imagine things as they will be in the future and as they have been in the past. They have an awareness of something else beyond our familiar day-to-day physical world. Human beings have discovered that there are things “unseen” that exist beyond the five physical senses of sight, sound, touch, taste, and smell (e.g. subatomic particles and certain forms of electromagnetic radiation).

Human beings are aware of their own mortality and understand that they have a limited amount of time on planet Earth. Most humans have a desire to be remembered favorably after they have died. The opinions of future generations matter to them, despite knowing that they are mortal. They are able to distinguish right from wrong, and to feel shame or guilt. The many aspects of the human spirit, both good and bad, are a part of our existence and experience that animals just do not possess.

There is an extra dimension to the character of human beings that is not present in any animals. For an animal, every day can be summarized as a struggle to survive and propagate. Many aspects of human behavior are also related to that need, but clearly many are not. The human desire to discover truths goes beyond the basic need for survival and has no counterpart in the animal kingdom. Human beings, unlike other living things, not only observe and interact with the world around them, but they also ask questions about it: Who, What, When, Where, Why, and How? They are aware of, and can enjoy the beauty of nature, such as a rainbow or sunset.

People's lives are not necessarily more difficult than animal's lives, but they are certainly more complex. In your neighborhood library, you will find rows and rows of biographies. Each biography contains thousands of words detailing the experiences in the life of one human being. Compare those experiences to the life of any animal on Earth, including the more complex such as primates. There is far more variety and detail within the experiences of one person's life than in the life of any one of the many different animals on Earth.

To illustrate the remarkable contrast between a human being's character and that of an animal, consider the following ten descriptive terms sometimes used when referring to human beings, but never used when describing animals:

- Deliberative—being able to discuss or carefully consider issues before deciding how to proceed.

- Poetic—going beyond simple communications and expressing things in the form of poetry.

- Lyrical—expressing things in a way that would be considered artistically beautiful or expressive.

- Inspired—doing exceptionally great things not normally expected of a human being.

- Contemplative— thinking carefully and studying things thoughtfully.

- Cultural—having an appreciation of the fine arts such as music, theater, or painting.

- Judgmental—tending to judge things harshly at times.

- Nostalgic—experiencing a yearning for the happiness felt in a former place or time.

- Prodigy—a young person recognized as having exceptional talents in some specific area.

- Genius—recognized as having a level of exceedingly rare or remarkable talent or intelligence.

There will never be a philosopher in the animal kingdom. There will never be a poet in the animal kingdom. There will never be a great composer or artist or musician or orator or debater or writer or storyteller or architect or chess master or teacher or medical doctor or lawyer or scientist or entertainer in the animal kingdom.

People enjoy keeping animals as pets. As pleasurable as it is to have a goldfish, rabbit, bird, turtle, cat, or dog in the home, none of our pets are ever associated with the terms descriptive of uniquely human abilities or talents. Our pets are never described as contemplative, nostalgic, deliberative, poetic, inspired, cultural, judgmental, or soulful, despite often being in the presence of humans.

In the one place in the universe where we know that life can exist and thrive, there is no natural reason to expect one form of life to be so extraordinarily different from ***all*** others. If there were not some kind of universal planning involved, one would expect that some animal species would exhibit at least a few of the human mental abilities and talents. Clearly, human beings are endowed with special, unusual, and unique qualities. The contrast between what it means to be human and what defines all other known forms of life is huge. Why are human beings singled out in this way if not for some kind of intentional plan?

Individual Uniqueness

Personality, mental ability, and natural talents are three important aspects of each individual human being, and of many individual animals as well.

Personality is determined by how a person tends to think, feel, and behave. Psychologists have concluded that an individual's personality is predominantly made up of a combination of five traits:

- Openness, how open a person is to experiencing and trying new things, includes attributes such as adventurous, curious, or creative.

- Conscientiousness, how dedicated and dependable a person is, includes attributes such as thoughtful, organized, detail-oriented, dependable, dedicated, disciplined, achievement-focused, or self-controlled.

- Extroversion, how comfortable a person is in the company of others, includes attributes such as talkative, sociable, assertive, or excitable.

- Agreeableness, how well a person gets along with other people, includes attributes such as helpful, kind, compassionate, or cooperative.

- Neuroticism, how emotionally stable a person is, includes attributes such as prone to anxiety, moodiness, irritability, worry, or having a low tolerance for stress.

Aspects of these personality traits can be seen in the behavior of animals, not just humans. Although individual animals from a particular species do show significant differences in personality traits, the same is not true of mental abilities and natural talents. Individuals belonging to a certain species of bird, fish, mammal, or other animal each have a similar set of mental abilities and natural talents. As with human beings, individual animals belonging to any given species have very similar physical structure to each other. However, unlike humans, they also have very similar mental abilities and natural talents. Each individual animal will therefore interact within their environment in a very similar way to others of their own species.

Individual humans, by contrast, have major differences in mental abilities and natural talents. For example, some people can create beautiful artwork, but they cannot carry a tune. Some people have a natural ability to solve complex puzzles; others can communicate easily in multiple languages. Each individual human being has remarkably different talents and abilities from all others.

Human Mental Abilities and Natural Talents

This unique characteristic of human beings became increasingly apparent during the Renaissance period from the 14th to the 17th century. “Renaissance” in Old French means "rebirth." The Renaissance was a period in history of cultural, artistic, political, and economic rebirth following the Middle Ages. The Middle Ages, also referred to as the Dark Ages, or medieval period, began with the fall of the Roman Empire, around the year 500, and lasted about one thousand years.

It is no coincidence that the Renaissance period began about the time the printing press became useful for the widespread sharing of ideas and knowledge. In 1440, Johannes Gutenberg, a German blacksmith, goldsmith, printer, and publisher, designed a printing press with “movable type printing” that allowed adjustable wooden or metal characters to be used along with an oil-based ink. Multiple lines could be printed at one time, dramatically reducing the time and cost of creating books that previously had to be individually handwritten. He revolutionized the creation of books and helped make the distribution of literature affordable.

Today, we are witnessing another renaissance of sorts: the further acceleration of idea-sharing made possible by the worldwide use of computers and the internet. I expect that this millennial renaissance will reveal even greater variations in talents and abilities among individual human beings.

A look at exceptional people from history illustrates how amazingly different individuals are from each other.

Leonardo da Vinci (15 April 1452 – 2 May 1519) was a multi-talented Italian with accomplishments in the areas of inventing, drawing, painting, sculpting, architecture, science, music, mathematics, engineering, literature, anatomy, geology, astronomy, botany, writing, history, and cartography. Best known as an artist, he is famous for his masterpieces such as the *Mona Lisa*, *The Last Supper*, and *The Annunciation*. He had no formal academic training, but he had an intense curiosity. His notebooks contain thirteen thousand pages of drawings, sketches, scientific diagrams, and thoughts on a wide variety of subjects. He described and sketched ideas for many practical inventions, but he died before being able to organize and publish them. Many of his ideas, such as those for manned flight, a practical parachute, an underwater diving suit, and a self-propelled cart (using springs for power) were not used until hundreds of years later.

Michelangelo (6 March 1475 – 18 February 1564), also born in Italy, was a sculptor, painter, architect, and poet considered by many to be the greatest artist of all time. His artistic works include the famous *Sistine Chapel* ceiling and the statue of *David*, a marble sculpture of the Biblical hero David. His works of poetry include over three hundred sonnets and madrigals. Much of his exceptional success as a sculptor and artist can be attributed to his detailed study of the human body and its proportions. By the age of 17, he was dissecting corpses at the local church graveyard to gain a deeper understanding of the human body. He had an extraordinary impact on art and society during the renaissance period and his influence continues to this day.

William Shakespeare (April 1564 – April 1616) was an English poet, playwright, and actor. His works, including numerous plays, sonnets, and other poetry, are known throughout the modern world. Many consider him to be the world's greatest dramatist. Exceptionally clever with words and images, Shakespeare wrote over one million words during his lifetime.

Galileo Galilei, born in 1564, was an exceptionally talented Italian scientist known for his work as an astronomer, physicist, engineer, philosopher, and mathematician. Albert Einstein called him the father of modern science [13]. He is believed to be the first person to invent a practical thermometer, by using the expansion and contraction of air in a bulb to move water in an attached tube. He built a useful version of a telescope that he used to help prove that the solar system was sun-centered. He also studied the craters and topography of the earth's moon, and sunspots. In 1610, he noticed three points of light near the planet Jupiter, at first believing them to be distant stars. Over time, he noticed that they appeared to move in the wrong direction relative to the background stars, and that they remained near Jupiter. He correctly concluded that they were not stars at all, but moons orbiting around Jupiter. Later that year, Galileo published his discoveries of Jupiter's satellites and other celestial observations in a book titled *Siderius Nuncius.*

Rembrandt (15 July 1606 – 4 October 1669) was a Dutch draftsman, painter, and printmaker. He produced hundreds of paintings, etchings, and drawings. He is best known for his portraits and for illustrations of Bible scenes. In addition to many portraits done for patrons, he produced over forty self-portraits.

Isaac Newton was an English mathematician, physicist, and astronomer. He developed his laws of motion and universal gravitation to describe the interaction between physical objects, and he explained the observed motion of planets around the sun. He made contributions in optics, including building the first practical reflecting telescope. He also contributed to the field of mathematics, and was a pioneer in developing calculus.

Wolfgang Mozart (27 January 1756 – 5 December 1791), a talented Austrian musician and prolific composer, began creating music at age five. In his short life, he produced more than 600 works of music, including 41 symphonies, 27 piano concertos, 5 violin concertos, 27 concert arias, 23 string quartets, 18 masses, and 22 operas.

Thomas Alva Edison (11 February 1847 – 18 October 1931) was an American inventor and businessman, considered by many to be America's greatest inventor. He developed many useful items in the fields of communication, sound recording, and motion pictures. Along with Nikola Tesla, he was one of the pioneers in the production and use of electrical energy. His inventions include the phonograph, the motion picture camera, and a long-lasting, practical electric light bulb. He held over one thousand U.S. patents, as well as patents in other countries.

Albert Einstein was a German-born theoretical physicist who expanded our understanding of the physical laws of matter, energy, and the fundamental forces of nature. He is best known for his theory of relativity, developed from two revolutionary ideas. One was that the laws of physics are the same for all observers in uniform motion relative to one another; the second was that the speed of light in a vacuum is the same for all such observers. His work proved the equivalence of mass and energy, and vastly improved our understanding of the elementary particles of nature and their interactions with the fundamental forces. He published more than three hundred scientific papers.

Not every human being has the same level of natural ability or talent in all areas. However, each person is remarkably different from all others and unique in their own special way. Despite having the same basic physical structure, their mental abilities and natural talents vary widely.

The differences in mental abilities and talents among individuals of any given species of bird, fish, or mammal have nowhere near the variation as do individual humans. Animal differences are almost entirely associated with personality differences.

Scientists have unsuccessfully attempted to connect differences in human physical makeup to explain the wide variation in individual mental abilities. For example, by studying human brains, scientists have tried but failed to find any correlation between intelligence and brain size. One study looked at the large brain size differences between two well-respected European writers. Ivan Turgenev (9 November 1818 – 3 September 1883) was found to have a brain that weighed more than 2000 grams. Anatole France (16 April 1844 – 12 October

1924) had a brain that weighed 1017 grams. Both men were successful poets and novelists. Anatole France was a French poet, journalist, and novelist with several best-sellers. He was a member of the Académie française and won the 1921 Nobel Prize in Literature. Ivan Turgenev was an admired Russian novelist, short story writer, poet, and playwright.

Genetics

Genetics research provides more insight into the physical makeup of living things. Every living organism has within its cells a genetic code that translates instructions specifying the physical structure of the organism. These instructions are contained within structures called chromosomes, made of genetic material called DNA (deoxyribonucleic acid). They are written in a coded language and passed from one generation to the next. The entire set of the components that make up an organism's DNA is called the genome.

An Oct. 25, 2011 article from the *Mobile DNA Journal* explains that the genomes of humans and chimpanzees are more than 98.5% identical at the protein-coding loci and this small difference is not sufficient to explain the extensive phenotypic differences between the two [14]. Human beings are in many ways physically similar to chimps, but functionally they are very different, especially in how they express themselves and interact with their environment and with others. It is obvious that there are incredible differences in their innate mental abilities and natural talents.

No one has ever observed a chimpanzee take time to gaze at the beauty of a sunset, rainbow, or spring wildflowers. You will never see two chimpanzees working together to accomplish a task such as building a shelter, carrying a load of supplies, or building a fire for warmth. Only human beings look at the wonders around them and ask themselves questions such as: "What are those pinpoints of light in the night sky?" and "What causes the moon to shine with a silvery light on some nights, yet have a reddish glow on other nights?"

Certainly, everyone would agree that there will never be a Leonardo da Vinci in the animal kingdom. There will never be a Shakespeare in the animal kingdom. There will never be an Isaac Newton. There will never be a Thomas Edison. There will never be an Albert Einstein. There will never be a Michelangelo or Rembrandt or Mozart either.

It seems amazing that chimp behavior is much closer to that of other mammals than to human beings, despite their genetic makeup being remarkably close to that of humans. This suggests that, although DNA is responsible for the physical similarity, the many other large differences involve something else. There is, in fact, an explanation for why chimpanzees lack so many human qualities while sharing so many similarities with other animal species. Something that human beings possess is missing from chimpanzees, and it exists outside of the physical makeup defined by the DNA.

The remarkable human qualities that are lacking in all animal species involve the use of certain mental processes not present in any other living things. Human beings possess a distinct and unique type of mental processing ability that allows humans to interact with their environment in a deliberative and controlled manner—so different from animals.

Evidence from the precise characteristics of the universe and from the behavior of living things shows that something extraordinary is involved in the functioning of the universe. Intentional planning explains why the universe has the physical laws, properties, matter, energy, and forms of life that it does. The next chapter provides further confirmation, by delving deeper into the mental process differences between human beings and all other living things. The recent development of computer systems provides this additional insight.

CHAPTER FIVE

COMPUTER SYSTEMS

"You know, if the hardware is the brain and the sinew of our products,
the software in them is their soul"
-Steve Jobs, microcomputer pioneer and co-founder of Apple Computer Inc.

Hardware and Software

Computer systems are made up of two very distinct but essential parts. ***Hardware***, designed and developed by computer hardware engineers, consists of the physical components of a computer system. ***Software,*** created by software developers and programmers, are sets of instructions that enable the hardware to process information and produce useful results. For example, software programs can instruct computer hardware to perform actions such as printing text, displaying images, or dialing a phone number.

Hardware provides the platform upon which the software operates. Hardware without functioning software is about as useful in this world as a paperweight. This is obvious from the over fifty million computer systems from the United States alone discarded and/or buried in landfills each year. When we include mobile devices, the total number of systems trashed or recycled yearly in the United States is over two hundred million [15].

Interestingly, software is not made of any material substance. It cannot be reduced to elements of matter and energy. Unlike computer hardware, it is not part of the physical world. It exists solely as a collection of instructions that can direct the operation of hardware components to produce an output result. If hardware is damaged or destroyed, the software still exists and will function once again if combined with other replacement hardware.

Before a novelist puts words on paper, the idea for the story exists in the writer's mind, independent of any physical reality. The information and events described in the story are not composed of physical matter and/or energy—unlike the paper and ink that allows the finished story to be shared with readers.

A music composer will have an idea for a new piece of music before it ever becomes a reality that we can experience. Only when combined with materials from the physical world —musical instruments or a singer's voice—can others enjoy the music.

Similarly, a computer software program first exists as thoughts in the mind of a software programmer. The non-physical software remains dormant and useless until combined with physical hardware. The hardware allows the information and instructions within the software to be used in a productive way. Software can only be functional and useful when combined with the appropriate hardware. In a sense, the presence of software is what brings the computer to life.

A company that produces thousands of computers of a particular model will produce each one with a nearly identical hardware design. These individual computers can each have different functions because despite having nearly identical hardware, the software installed on each can be very different.

A computer has two distinct types of software. The ***operating system software*** determines how the computer handles the basic input, output, and other routine internal functions of the system. Every different computer model type has its own unique operating system software. The other type of software, ***application software***, determines the ultimate functional capability of the computer. It allows the computer to interact with the outside world by accepting input information and then producing useful output results. Unlike the operating system software, the application software can be significantly different on individual computers of the same model. On small portable computers, such as tablets or smartphones, different types of application software are commonly called "apps."

The physical hardware, operating system software, and application software make up the three essential parts of a functioning computer system. Two computers will have the same hardware and operating system, but may have quite different application software installed. The application software is what makes each individual computer system unique.

For example, two friends may each have a personal computer with nearly identical hardware and the same operating system software installed. However, the two computers may have completely different uses. One may contain software designed to work with arts and crafts projects and to manipulate photos and graphics. The other computer might have software programs installed that are geared more towards processing numbers, such as evaluating data from experiments or for financial analysis. One computer can handle numbers well, and the other, with essentially the same hardware, will be useful for more artistic pursuits.

Application software comes in two forms. ***Automatic application software*** requires no user input once set up and usually processes rapidly, so it's useful when time is important. Examples of these processes are found in software designed for automated factory robots and in programmable thermostats that automatically adjust the temperature as needed.

The second form, ***controlled application software***, is the type most familiar to people. The input, and therefore the resultant output, is under the direct control of the user. Using controlled application software is more time-consuming than the automatic software, but is necessary for applications such as word processing, spreadsheets, publishing, databases, internet browsing, and communications.

Comparison of Computers and Living Things

By using a familiar example, an analogy can help explain or clarify concepts and ideas by comparing the shared qualities of two different things. Analogies enable us to relate past experiences to new situations. An analogy brings clarity. In previous chapters, analogies have helped to explain important concepts supporting the reality of our well-planned universe:

- The expanding universe, where galaxies are constantly moving away from each other, becomes clear from the comparison with a loaf of raisin bread baking in the oven, where the distance between each raisin in the bread steadily increases as the bread rises.

- The formation of molecules—structures consisting of two or more atoms bound together—is better understood by comparing them with Lego(TM) building blocks that are also combined to produce complex structures much larger than themselves.

- The example of a well-planned garden discovered in an isolated mountain meadow shows how evidence from recent scientific discoveries now enables us to recognize our well-planned universe. A variety of different types of evidence, brought together, provided overwhelming confirmation that the hikers had found an intentionally constructed garden. Similarly, the precise nature of so many aspects of the universe confirms that what we observe is the result of a deliberately planned and intentional act.

As with these analogies, a comparison between computers and living things will make clear how non-physical processes enable living things to function much like non-physical software functions with computer hardware. A computer is useful only after software is combined with hardware. When added to the system, application software can be considered the "mind" of the computer in the sense that it allows the computer to process information and produce useful results.

As with computers, all living things can successfully interact with the outside world only when the physical hardware part and the non-physical processing part work together properly. A human's or animal's "body hardware" is like the hardware components of computer systems. Individual humans and animals can be viewed as a type of "organic computer" in the sense that they are made up of physical and non-physical parts. Both parts are needed, but it is the processing part that accepts input from the surrounding environment and makes it possible to interact with the outside world.

The basic hardware components of a simple computer system are shown in Figure 2.

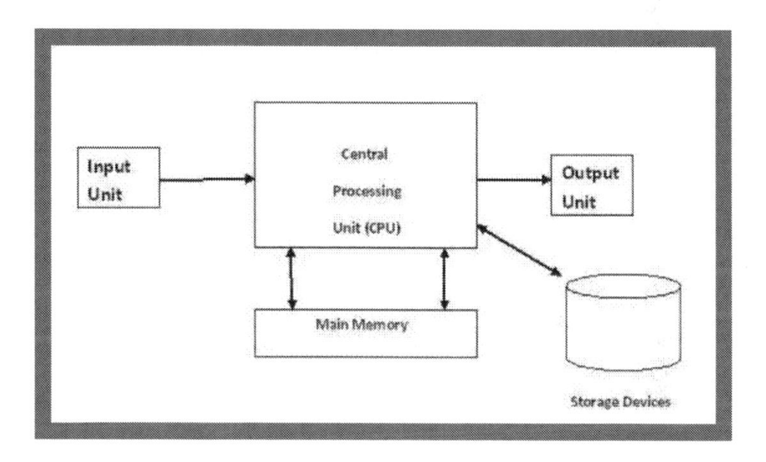

Figure 2. Basic hardware components of a simple computer system: Input device, System memory, Central Processing Unit, Storage devices, and Output device (e.g. printer or display)

A manufacturer can produce thousands of nearly identical computers. The major difference between them will be determined by the type of software installed on each. Similarly, the physical body hardware of every individual human being has a nearly identical structure. A single book on anatomy can describe in detail the physical makeup of every human body on Earth.

In the same way that all units of any given computer model have essentially identical hardware, individual humans, as well as individuals of any species of animal, are physically much the same as each other. In other words, our physical body hardware components have structures that are essentially the same for all of us.

Consider the twelve major physical "hardware" systems of the human body:

- The skeletal system consists of 206 bones connected by a network of tendons, ligaments, and cartilage. Together, they support the body and protect the body's organs.

- The nervous system consists of cells that control certain internal functions of the body and the body's responses to external stimuli. The human nervous system consists of the central nervous system (CNS) and the peripheral nervous system (PNS). The CNS consists of the brain and spinal cord; the PNS is a network of nerves that carry impulses to and from the CNS. The PNS functions like the body's electrical wiring, transmitting signals between different parts of the body.

- The cardiovascular system uses arteries, veins, and capillaries to circulate blood throughout the body, transporting and delivering oxygen and nutrients to cells and carrying their waste products away.

- The integumentary system consists of skin, hair, nails, and glands that help to regulate body temperature and protect the body from damage caused by loss of water or harm from the outside environment.

- The endocrine system is made up of various glands that produce hormones to provide chemical communications within the body. These hormones regulate body functions such as growth, metabolism, respiration, reproduction, sensory perception, and sexual development.

- The muscular system consists of 650 muscles that not only support movement such as walking, talking, sitting, standing, and eating, but also help to maintain posture and to circulate blood and other substances throughout the body.

- The lymphatic system is a network of tissues and organs that helps eliminate toxins, waste, and other unwanted materials from the body. A major function of the lymphatic system is to deliver infection-fighting white blood cells throughout the body.

- The respiratory system exchanges oxygen and carbon dioxide with the environment. The lungs work with the circulatory system to supply oxygen-rich blood to all the cells in the body. The blood then collects carbon dioxide and other waste products to transport them back to the lungs, where they are expelled out of the body through exhalation.

- The excretory and urinary system eliminates waste from the body. It includes the liver, sweat glands, kidneys, and bladder. It also helps to maintain the balance of chemicals and water in the body.

- The reproductive system consists of the organs required to produce offspring.

- The digestive system processes nutrients for the body. The gastrointestinal tract consists primarily of the mouth, esophagus, stomach, small intestine, and large intestine (colon). The liver, pancreas, and gallbladder are also part of the digestive system.

- The immune system defends the body against disease. Bone marrow produces white blood cells that attack viruses and harmful bacteria. The spleen destroys old or damaged red blood cells and makes disease-fighting components of the immune system, including antibodies. Lymph nodes supply and store cells, such as the white blood cells that fight infection and disease.

As with computer hardware, the twelve human body hardware components require an operating system to function. When working properly, no deliberate conscious control or outside intervention is necessary. Some well-known examples are the processes involved in the digestion of food, maintaining internal body temperature, regulating heart rate, and respiration. The human internal operating system functions from within the central nervous system (CNS), much like a computer operating system functions from within a central processing unit (CPU) as shown in Figure 2.

Like computer models, each different life form has a distinct physical makeup, and each requires its own specific type of operating system to function properly. Not all animal species have the same type of central nervous system as humans. However, all animal species do have an operating system specific to their species that makes it possible for various parts of their physical body hardware to function without conscious control.

Automatic and Controlled Central Nervous System Processes

As described previously, computers use operating system software and two types of application software—automatic application software and controlled application software. Similarly, human beings use automatic and controlled CNS processes to interact with their environment. Automatic CNS processes have no need to be consciously controlled. We instantly—automatically—react by pulling our hand away when we accidentally touch a hot stove. Controlled CNS processes, however, require time-consuming, deliberate thought, such as when we make a decision or create a work of art.

I will use the term "***life software***" when referring to the processes used by living things to clearly distinguish them from processes performed by computer software. Accordingly, the three types of life software that enable living things to function are ***body hardware operating system software, automatic life software,*** and ***controlled life software***.

Automatic Life Software

Animals and human beings have automatic life software, including instinct, intuition, and emotion, that affect behavior as they interact with their surroundings. Sensory input received from physical senses—sight, smell, hearing, taste, and touch—can be quickly processed to determine the appropriate action. This is similar to the way a computer uses automatic application software to quickly produce an appropriate output, such as when a programmable thermostat automatically adjusts temperature. Automatic life software is extremely effective in ensuring survival.

Some of these automatic processes are instinctive, and others are learned through experience, such as learning to walk, learning not to touch the surface of a hot stove, and developing balance while riding a bicycle. Once learned, they become automatic and normally don't require deliberative thought. However, if we encounter an unexpected situation while walking or riding a bicycle, we may need to use our more time-consuming controlled processes to decide how to handle the situation.

Animals and humans use automatic life software when they encounter a sudden threat and react in one of three possible ways—fight, flight, or freeze. They may immediately flee to escape the situation, as a rabbit might do. When escape is not an option, some will stand their ground, confront the threat, and prepare to fight. Some predatory mammals, such as grizzly bears, will not hesitate to attack when they feel threatened. Other animals will attempt to avoid a confrontation by ceasing all movement. The best-known example is the opossum that has the famous characteristic of seizing up, falling on the ground, and appearing to be dead. The opossum has no control over this behavior of "playing dead." This involuntary reaction is triggered by stress and can last for hours. These automatic fear responses are a type of automatic life software and are effective in helping the individual avoid injury or even death.

Many animals have amazing natural instincts and intuitions that make up part of their automatic life software. Some of the most astonishing are seen in the behavior of migratory species.

The annual migration of the Arctic tern, from the high Arctic breeding grounds to the Antarctic ocean, is believed to be the longest seasonal movement of any animal. During their migration, these birds have been tracked for more than 80,000 km, over deep water, and at a considerable distance from continental shelf margins [16]. To prepare for their difficult migration journey, the birds somehow know that they must bulk up on food in the preceding few weeks to store fat for energy on their long trip. They manage to find their way each season, but how they navigate is not fully known. Scientists believe that the birds can get compass information from the sun, the stars, and by sensing the earth's magnetic field.

Equally fantastic is the migration journey of some butterflies. Possibly the most spectacular example, Monarch butterflies, travel in the fall from Canada and the northern United States to overwintering sites in central Mexico. They start the return trip in March and usually arrive back north in July. No individual butterfly completes the entire round trip, as the female monarchs lay eggs for the next generation during their northward migration. The mechanism for this behavior is unknown, but some have proposed that a type of genetic memory must be inherited in each generation to make this possible.

Another astonishing example of programmed-like behavior comes from the study of sea turtle migration. Sea turtles navigate the open sea for hundreds, sometimes thousands, of kilometers. Years after their birth, adult females return faithfully to nest and lay their eggs on the same stretch of coastline where they originally hatched. Sea turtle navigation has been studied for decades, but is still not understood. The most likely explanation is that they somehow use Earth's magnetic field to help guide them at sea. The turtles appear to have a type of inherited "magnetic map" with instructions that direct them as they move across the ocean.

One last example of amazing programmed-like behavior is that of an ant colony. As a child in elementary school, I saw ants as my first pets. Ants, in my opinion, seemed ideally suited to be pets. My mother, however, considered them pests. "Maybe a goldfish would be better," she suggested. I briefly considered the idea before redirecting my attention back to the more interesting activities of the ant society. Mom reluctantly agreed to allow ants in the house, but only if they were properly contained. After school, I would venture out behind our house in southern California and collect some of the large red ants that made their home at the edge of the back alley. I would poke at the ground with a stick, then scoop up ants into a large glass jar as they came out to protect their nest.

I recall that one day, two concerned police officers drove past me in the alleyway and asked what I was doing. “Catching ants,” I replied. They asked where I lived and I pointed across the alley to the house. That seemed to satisfy them, so I continued with my project as they drove off.

The ants were not happy as I disturbed their underground city but I just had to know what they were doing down there. I would collect a few dozen of them in a glass jar containing sandy soil, then watch them dig new tunnels. I learned to cover the outside of the jar with dark paper to keep out daylight. The ants would then dig tunnels along the side of the jar, and I could view their underground activity by occasionally removing the paper. I fed them small pieces of bread soaked in sugar water.

I remember watching with fascination as individual ants carried on with the daily duties necessary to sustain the colony. By observing the lowly ant, a child can learn much about how a society operates, and how individuals can accomplish substantially more by working together than by working separately.

I felt like the master of their little universe. It would have been even more interesting if I could have given them the ability to individually make decisions and plan what they would do next. That superpower I did not have. They had no idea, nor did they care, who I was or why I was there. They never acknowledged my presence as long as I did not disturb them. They each instinctively had their own special job to do and that kept them busy. They seemed content with their lot in life.

Each ant has a distinct role to play. How is it that such tiny creatures, each produced by the same queen ant, have such different duties within the colony? Some forage for food, some defend the colony from invaders, and some tend the young ants until they mature. Each has the physical hardware and automatic life software necessary for them to carry out their distinct daily tasks. The ants behave like little robots, each programmed for a specific role within the colony.

There are thousands of different species of ants, each specialized in their behavior and how they ensure the colony's survival. For example, leafcutter ants remove leaves and carry them back to the nest. In chambers within the nest, they grow a fungus on the leaves for their nourishment. Certain ants within the colony have the job of protecting the leaves from parasitic flies and wasps.

Several unusual species of ants, commonly known as army

ants, live in temporary colonies. When food supplies become short, they leave the nest and travel great distances in long columns. Large soldier ants flank the sides of the columns, protecting the smaller ants within the column. They form bridges and tunnels with their bodies to aid their progress. Along the way, they capture insects, spiders, and even small vertebrates such as frogs, snakes, and lizards.

Even a tiny creature like an ant or a bolas spider appears to have programmed complex task behavior that benefits survival. This behavior, although more complex, is remarkably like that of a computerized factory robot that has been programmed to perform a specific task. An individual animal is much like a computer device, such as a smartphone, that is designed for specific tasks. Its hardware and software components are replicated by the thousands—or millions—each one nearly the same as others of the same "model."

Consider the difference between a computer and an old-style television set. Each is designed for specific tasks and is useful in its own way. There are many physical similarities between a computer and a television set. The physical parts that make up the hardware, such as metal, plastic, and silicone, are similar in both. However, the computer is functionally more capable than an old-style television set. The computer, unlike the television, is able to make use of the non-physical intelligent application software.

In the same way, there are many physical similarities between human beings and certain animals, especially mammals, but functionally they are quite different. This suggests a non-physical, uniquely human cognitive element. Human beings possess deliberative thought and complex decision-making abilities that are absent in animals. In addition to the automatic life software necessary for long-term survival, human beings have the ***controlled*** life software not available to animals.

Controlled Mental Processes

Human beings are dramatically different in many ways from the many animal species found on Earth. DNA analysis has revealed that the physical makeup of the body hardware of the chimpanzee is very close to that of a human being. The astounding difference between the two is found in the ability of human beings to use controlled life software within the brain hardware. For clarification, instead of using the term "controlled life software," I will refer to this uniquely human quality as "***controlled mental processes,***" since this ability comes solely from the unique mental processes that occur within the human brain.

Examples of some of these controlled mental processes are:

- Planning or getting ready for an upcoming event.
- Deciding what item to purchase using a gift card.
- Thinking about the consequences before making an important decision.
- Listing the pros and cons before deciding to buy or rent a home.

- Researching possible choices before deciding where to go on vacation.
- Researching available makes and models before purchasing a new vehicle.
- Deciding what career to pursue.
- Deciding whether or not to propose marriage or accept a marriage proposal.
- Deciding whether or not to move to a different city or country.
- Studying for an upcoming school exam.
- Creating a work of art such as a drawing, painting, or sculpture.
- Creating poetry or writing a book.
- Composing a piece of music.

Human beings are so different from all other forms of life on Earth because they possess and use both automatic life software and controlled mental processes. The time-consuming controlled mental processes enable human beings to study, create, and make decisions. Animals are entirely motivated by survival and propagation. Their extremely effective and fast-acting automatic life software is well-suited for those two purposes. In fact, an animal's behavior is fully determined by its automatic life software. Animals operate within their environment using intuition, instinct, emotion, and sometimes learned or trained behavior.

When faced with the difficult question, "Why are humans so different from animals in their thinking and mental abilities?" researchers invariably proceed to answer the much easier question, "What similarities can be found between humans and animals?" Most studies of animal behavior focus so much on studying the similarities between human and animal behavior that the significance and magnitude of the differences between humans and animals is overlooked. These studies show some interesting similarities between the automatic life software of humans and animals, but they are unable to explain the remarkable controlled mental processes unique to human beings.

An important distinguishing characteristic of those living things that lack the benefit of controlled mental processes is the element of choice. Animals do not have a choice in their behavior. Each of them must do what intuition, instinct, emotion, and/or their learned behavior tells them to do.

The presence of uniquely-human life software enables humans to make thoughtful decisions through controlled mental processes. Our physical body hardware is in numerous ways inferior to many animal species. The controlled mental processes make us who we are as individuals, resulting in each of us possessing different talents and abilities. Unlike animals, each person has a unique set of controlled mental processes that allows them to interact with the world in a deliberative and controlled manner.

The results produced from a smartphone app depend upon the input. By choosing what information to input, we can control our own life software in the way someone controls the "app software" in their smartphone. As with so many things, the quality of what goes in determines the quality of what comes out. "Garbage in, garbage out," or GIGO, is a well-known computer software adage.

Many animals have larger brains and nervous systems than humans, yet human understanding and experience encompasses so much more than any animal's experience. As explained previously, the number of brain or nerve cells appears to have little correlation with an animal's or a person's mental abilities.

In human beings and some animals, automatic life software operates within the entire nervous system, including their brain. Controlled mental processes, unique to human beings, function from within the human brain.

The internal, automatic, and controlled functions of computers, humans and animals is shown in the following chart.

Computer Software and Life Software

	Computers	Human Beings	Animal Species
Internal Functions	Operating System Software	Body Operating Life software	Body Operating Life software
Automatic Functions	Automatic Application Software	Automatic Life software	Automatic Life software
Controlled Functions	Controlled Application Software	Controlled Mental Processes	Absent in Animals

Chart 1. Internal, Automatic and Controlled Functions

The FlyEM project at the Howard Hughes Medical Institute's Janelia Research Campus in Virginia has analyzed a portion of the brain of the common fruit fly, Drosophila melanogaster, and produced some fascinating results. In 2020, the project team, in collaboration with Google(TM), produced a three-dimensional image of the connections (circuitry) between nerve cells (neurons) in a large fraction of the fruit fly brain. The electron microscope image includes areas of the brain involved in key functions, such as learning, smell, and spatial navigation. A portion of these results is shown in Figure 3. Once again, the similarity between the body hardware of a living thing and computer system hardware is evident.

Dozens of researchers from various disciplines including biology, computer science, and physics were involved in the project. The research continues and the team plans on analyzing the circuitry to produce a map of the entire Drosophila nervous system. If you are interested in investigating the results further, the team is making the data publicly available. This "wiring-diagram" will enable scientists to better understand how the fly's nervous system functions [17].

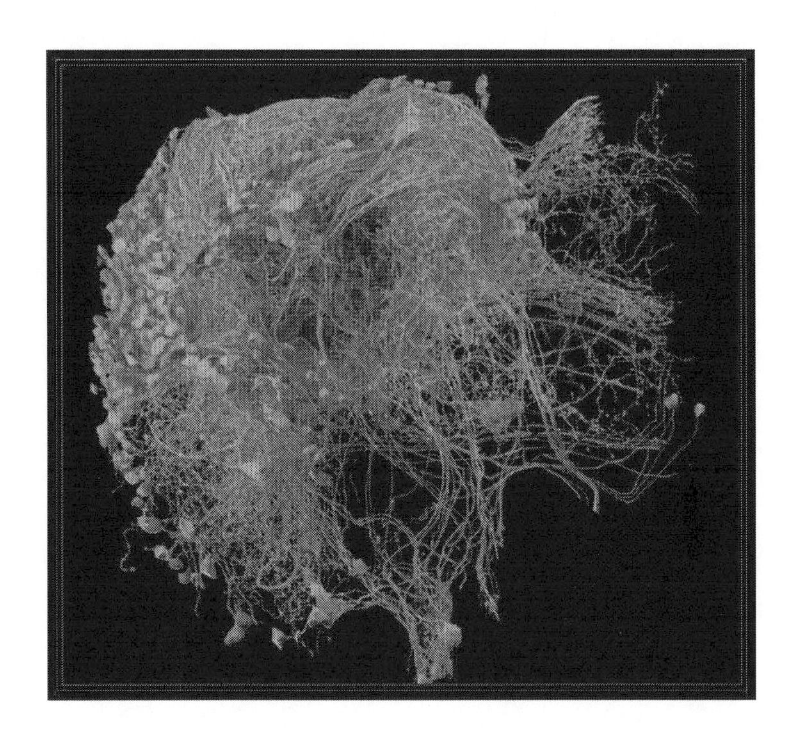

Figure 3. Connections between nerve cells in part of a fruit fly brain

Materialism and Mind-Body Dualism

If one considers the physical brain as part of a person's physical body "hardware," it is reasonable to consider the operation of the mind as part of the non-physical life software. The relationship between mind and body has two common viewpoints which have been hotly debated throughout history. Materialism holds that our minds are inseparable from our bodies, and that all reality is the result of material interactions. It suggests that mental states and the operation of our minds are solely the results of material interactions. The alternative point of view was described by the famous 17th-century French physicist, mathematician, and philosopher René Descartes (31 March 1596 – 11 February 1650), the "Father of Modern Philosophy." Descartes is also the inventor of the Cartesian coordinate system and the founder of analytic geometry.

Descartes wrote a treatise, "Passions of the Soul and The Description of the Human Body," suggesting that the body has material properties and works like a machine. He described the soul as non-material and therefore not following the same laws of nature as the physical world. Of course, the concept of the digital computer system was not available to Descartes in the 17th century. Descartes' "mind-body dualism" point of view holds that mind and body are two completely distinct and separate things, that the mind is different and independent from the physical brain. The "hardware-software dualism" present in all digital computer systems and the concept of human life software helps to explain Descartes' "mind-body" rationale. The advent of the computer has brought clarity to Descartes' insight from so many centuries ago.

You may be familiar with Descartes' famous dictum, "I think, therefore I am." I am reminded of the famous French sculptor François-Auguste-René Rodin (12 November 1840 – 17 November 1917) who produced the well-known bronze sculpture "The Thinker," circa 1880, illustrated in Figure 4.

Figure 4. Illustration of Rodin's bronze sculpture,
"The Thinker," circa 1880

"I think; therefore I am." "I speak; therefore I am human." "I read; therefore I am human." "I write; therefore I am human." "I create; therefore I am human." "I build fire; therefore I am human." "I ask questions; therefore I am human." "I design structures and artwork; therefore I am human."

Steve Jobs (24 February 1955 – 5 October 2011) was a pioneer in the development of the microcomputer during the 1970s and 1980s. As one of the founders of Apple Computers, he was instrumental in launching innovative products that revolutionized consumer technology, including the iPod, the iPhone, and the iPad (trademark Apple Inc.). At the Apple Worldwide Development Conference in June of 2011, Steve Jobs said of Apple Computers, "You know, if the hardware is the brain and the sinew of our products, the software in them is their soul [18]."

To my knowledge, Steve Jobs was not aware that his words could be significant to the understanding of the essence of what it means to be human. He died later that same year on October 5, 2011. The morning before his death, Jobs called his sister Mona Simpson and asked her to come to Palo Alto. In her eulogy, Mona reported that shortly before his death, Steve looked at his loved ones, then looked over their shoulders and spoke six short words. Steve Job's final words were "OH WOW. OH WOW. OH WOW [19]."

The concept of human life software and controlled mental processes makes it possible to understand why there will never be a philosopher or poet in the animal kingdom—or a great composer or musician or artist or orator or debater or writer or storyteller or architect or chess master or teacher or medical doctor or lawyer or scientist or entertainer.

Human beings are not only extraordinarily different from all animals, but remarkably different from each other. This huge diversity in the working of the controlled mental processes, despite what is essentially the same physical structure in all humans, is truly amazing. Each person has his or her unique human life software. This helps to make sense of the many examples of exceptional individuals such as da Vinci, Michelangelo, Shakespeare, Galileo, Rembrandt, Newton, Mozart, Edison, and Einstein. Each of them was endowed with remarkable talents and abilities, but in many different areas of the arts and in science.

Child Prodigies

Some children under the age of ten have demonstrated remarkable abilities in art, mathematics, or music, long before any formal training. When we look at controlled mental processes, it helps us to understand these child prodigies. Some well-known examples are listed below.

Blaise Pascal (19 June 1623 – 19 August 1662) was a French mathematician, physicist, inventor, and writer. As a child, he developed a deep interest in mathematics and science. He had no formal education, but received his primary education from his father. At age thirteen, Pascal and his father began attending discussions in Paris with a group of scientists and mathematicians, including René Descartes (see previous discussion of Descartes). At age sixteen, he wrote a significant treatise about projective geometry. While still a teenager, he began to develop mechanical calculating machines. His arithmetic machine, the Pascaline, was the first practical and widely used calculator.

Sor Juana Inés de la Cruz (12 November 1648 – 17 April 1695) was a scholar, philosopher, composer, and poet. As a female born in the 17th century, she had little access to formal education and was almost entirely self-taught. It is believed that she learned to read by age three. She produced numerous dramatic, comedic, and scholarly works, including many plays and poems.

Wolfgang Amadeus Mozart (27 January 1756 – 5 December 1791) was a musician and composer. At age five, he began creating music and was already playing songs on multiple instruments, including the harpsichord. He began playing in public at the age of six. At age nine he wrote his first symphony, and by age twelve had produced ten symphonies.

John von Neumann (28 December 1903 – 8 February 1957) was a Hungarian-American mathematician, physicist, and computer scientist. At age six he could divide two eight-digit numbers in his head and converse in Ancient Greek. By the age of eight, he was familiar with differential and integral calculus.

Stevland Hardaway Morris (born May 13, 1950), better known by his stage name Stevie Wonder, is an American singer, songwriter, musician, and record producer. By age nine he had taught himself how to play the piano, drums, and harmonica. By age twelve he had released his first album.

Alma Elizabeth Deutscher (born February 2005) is an English composer, pianist, and violinist. At age six she composed her first piano sonata; at age seven she completed a short opera,

"The Sweeper of Dreams"; at age nine she wrote a concerto for violin and orchestra; at age ten she wrote her first full-length opera; at twelve, she premiered her first piano concerto.

Hardware Eventually Fails;

Software Endures

Computer hardware will eventually wear out and fail to function properly. The natural order of all physical systems is to decay over time. Computer software, however, can be upgraded or easily replaced, so updated versions of software actually improve, rather than decay. Software updates have become commonplace with most computer applications.

Computer software can also be permanently preserved with a computer backup, whereby an exact copy of the software is made and stored away to be used later if needed. Backups are saved and stored on local computer systems, through uploading to remote locations commonly known as "The Cloud." The Cloud consists of computers that can be remotely accessed, using the internet to store, manage, and process data. If computer hardware is damaged, a duplicate of the original computer system can be produced by installing the backup software onto a new computer system of the same model type.

When computer hardware is damaged or wears out, the software cannot function properly and the hardware must be repaired if possible. If a critical piece of hardware cannot be repaired or replaced, the computer will have to be discarded. In that case, it will most likely end up buried in a landfill.

Similarly, when our physical body hardware wears out or is damaged, our ability to function will be diminished. For example, the decline in the condition of our physical brain with age will hinder the functioning of our mind. Life software, being an independent non-material entity (like computer software), is not subject to the same process of decay that affects the brain and the other parts of the physical body. However, it does require healthy body hardware to function properly. When a critical body part, such as an essential body organ, like heart, lungs, or liver fails, and cannot be medically repaired or replaced, the body will cease to operate. The life software will no longer be able to function within the body hardware.

A few years ago, I personally experienced a health issue that revealed to me how a problem with brain hardware can affect the ability of a person's mind to function properly. Fortunately, my body hardware problem was not permanent; it was corrected, and my life software functioned normally again.

Over several months, I had noticed some difficulty with concentration, and a slowing of my overall thought processes. I was fatigued and constantly thirsty, waking up several times during the night with intense thirst and a dry mouth. A blood test showed that I had hyponatremia, low levels of sodium in my blood.

It is often difficult to determine the underlying cause of this condition because it can be the result of heart disease, lung disease, intracranial disease, kidney disease, various types of cancer, hormone problems, or taking certain drugs. I had many different tests including multiple X-ray images, CT (computed tomography) scans, and MRI (magnetic resonance imaging) scans. I also had regular blood tests over several months to monitor the condition.

I was finally diagnosed with an abnormal increase in a hormone that causes fluids to be retained in the body's cells. Normally, this hormone signals the body to maintain the correct fluid content in the body cells. In my case, the continued abnormal release of this “antidiuretic hormone” resulted in an intense thirst and a buildup of fluids in the cells of the body. Unless brought under control, the continued buildup of fluid in brain cells leads to a buildup of pressure, causing mental difficulties. Eventually, this can lead to confusion, seizures, and even coma. The only effective long-term treatment for controlling the hormone is to find and treat the underlying cause.

I saw thirteen different doctors and numerous other medical professionals over a period of several months. The symptoms continued to worsen, and the situation may have deteriorated further had it not been for one kidney specialist. Fortunately, that doctor suggested that I see a sleep specialist because of my difficulty with sleeping through the night. A sleep test showed that I had severe sleep apnea. I received a prescription for a CPAP (continuous positive airway pressure) machine to supply sufficient air pressure at night to keep my airway open and to allow easy breathing during sleep. My intense thirst went away after a few days, and my blood sodium level began to slowly rise. It took several weeks of adjusting before I could sleep soundly through the night, but my symptoms gradually improved during that time. Today, my blood sodium levels are back to normal and I feel well again. My difficulty with sleeping was not a symptom of the medical condition, but the cause of the problem. When my body hardware problem was resolved, my mental processes functioned normally again. This was a personal example of how the mind's life software can function normally again if the physical brain hardware is properly restored.[2]

Investigations into the cause of dementia help us to understand the relationship between the mind and the brain. Dementia describes a decline in a person's mental acuity, resulting in difficulty with thinking, remembering, and reasoning skills. Dementia often develops gradually and is more common in elderly persons. Most examples of the condition come from Alzheimer's patients, but other illnesses, such as Parkinson's disease and Lewy body disease, can also lead to dementia. Physical damage to nerve cells and their connections in the brain, most often the result of a buildup of proteins, damage the nerves' internal structure. CT and MRI scans reveal the structure of the brain and can show the loss of brain mass associated with Alzheimer's disease and similar conditions.

---

[2] Most people know someone who has sleep apnea to some degree. In most cases, sleep apnea is mild and there is no need for concern. I had no idea that something as common as sleep apnea could become severe enough to cause such serious health issues. If you learn nothing else from this book (and I sincerely hope that is not the case) at the very least I hope that you will now make sure to get a good night's sleep.

My father died in 2003, a few years after being diagnosed with Lewy body dementia. His experience was an example to me of how brain hardware health can affect someone's ability to use controlled mental processes. When I would visit him at the assisted living facility, I could tell that he was still the same person that I had known all my life. But despite appearing physically healthy, he was unable to fully express his thoughts.

The well-known and admired actor and comedian Robin Williams (21 July 1951 – 11 August 2014) also had Lewy body dementia. In an *American Academy of Neurology* publication, Robin William's wife, Susan Schneider Williams, wrote "Robin is and will always be a larger-than-life spirit who was inside the body of a normal man with a human brain ... He kept saying, 'I just want to reboot my brain.' ... My husband was trapped in the twisted architecture of his neurons and no matter what I did I could not pull him out [20]."

Recent research has shown that people's subjective age (the way they feel inside) is often very different than their age measured in number of years. Many older people feel much younger mentally than you might expect. Using MRI scans, researchers have found that elderly people who describe themselves as feeling young mentally show fewer signs of physical brain aging when compared with others who feel their chronological age.

An interesting study, done by the Departments of Psychology, Seoul National University, and Yonsei University in Seoul, South Korea, was published in the open-access journal *Frontiers in Aging Neuroscience*. Scientists performed MRI brain scans in sixty-eight healthy people between the ages of

59-84, looking at gray matter volumes in various brain regions. Participants completed a survey that asked whether they felt older or younger than their age, and other questions assessing their cognitive abilities. On average, people who felt younger than their age scored higher on the memory test and were less likely to report depressive symptoms. The scans for those who felt younger than their age showed increased gray matter volume in key brain regions. In an interview, Dr. Chey from Seoul National University stated that, "We found that people who feel younger have the structural characteristics of a younger brain [21]."

The MRI imaging showed that the participants who felt younger had the brain hardware characteristics of a younger, healthier brain. The life software of the mind continues to function well into old age unless the physical brain health declines. This is consistent with the fact that the life software of the mind is independent from the physical brain. Once again, one is reminded of computer software, which does not deteriorate and remains functional unless the computer hardware wears out or is damaged.

CHAPTER SIX

REALITY REVEALED

"What the human being is best at doing is interpreting all new information so that their prior conclusions remain intact."
Warren Buffett, business magnate

An Abundance of Evidence

Evidence from the laws and properties of the universe, along with the many examples of uniquely human qualities, clearly points to a planned universe. Further independent confirmation comes from the study of computer systems and the controlled mental processes that are present in human beings, but absent in animals. Now that we have collected and reviewed the available information, we can arrive at an explanation for the nature and origin of our world and the universe beyond.

Despite initial doubts, skepticism, or uncertainty, a correct conclusion is possible if there is sufficient relevant evidence. Evidence of an intentional act can sometimes be obvious when we look at observations that show up at a later time. When presented with a theory or claim, an inquiring skeptic will look for such evidence.

The example of a cultivated garden in an isolated mountain meadow shows how a variety of evidence can collectively lead to the right conclusion. After careful consideration, the hikers confidently concluded that they had come across a deliberately planned garden plot.

To prove a theory or claim, an investigation should include both direct and indirect evidence from each of the five main types of evidence—physical, documentary, testimonial, demonstrative, and scientific. Examining evidence from multiple independent sources, while carefully avoiding biases, will almost always lead to a correct conclusion.

History shows that major advances in science frequently result from examining new evidence that initially appears to conflict with existing opinions. Recent scientific discoveries have led to a new, deeper understanding of the fundamental nature of the universe.

Everything detected in the universe is made up of matter or energy or a combination of both. Modern physics has shown us that matter and energy can be converted from one to the other under certain conditions.

The physical laws and properties of the universe control the interaction of matter and energy in a very specific way, allowing galaxies, stars, and planets—as well as life—to not only exist but persist over time. Because the universe maintains this stability, the universe is not an empty, lifeless void. Instead, the universe has exactly the right qualities for life to thrive on at least one planet.

In the last century, scientists discovered that the universe is expanding at a rate sufficient to prevent everything in the universe from collapsing due to the force of gravity. This is yet another fortuitous aspect of the universe.

Four fundamental interactions—gravity, strong nuclear, weak nuclear, and electromagnetic—affect matter and energy. The interaction of these forces allows the process of nuclear fusion to occur; this is how radiant energy from stars such as the sun is possible. If not for the precise nature and extraordinary strength of the strong nuclear force, two or more protons could not exist in the nucleus of an atom, and the many stable chemical elements would not exist. Massive objects such as stars and planets exist only because of the precise nature and strength of gravity.

The simplest form of a universe would consist of randomly distributed matter and/or energy, but little else. An architect or planner would be unlikely to devise such an uninteresting universe. When matter and energy interact and change form, variety within the universe is possible, and things become interesting. In a planned universe, the architect would incorporate interesting, complex features, rather than create the simplest possible design.

The universe is structured in a way that allows a large number of different types of matter, energy, and interactions to not only exist but persist over time. Amazingly, the universe also happens to contain a huge variety of lifeforms on at least one planet that we know of. The most fascinating and intriguing form of life would have the ability to make conscious choices and to design and create things.

Despite so much evidence in favor of intentional planning, and the lack of evidence for an alternative, some people still refuse to accept the idea that humankind occupies a special place in existence. However, the human race stands out as dramatically different from the more than seven million species of animals on Earth.

Human beings can think about what others are thinking. They are able to escape from their present point of view. They can imagine things as they will be in the future and as they have been in the past. They have an awareness of something else beyond our familiar day-to-day physical world. Human beings have discovered that there are things "unseen" that exist beyond the five physical senses of sight, sound, touch, taste, and smell (e.g. subatomic particles and certain forms of electromagnetic radiation). They can imagine things as they will be in the future and as they have been in the past. They can understand the concept of cause and effect.

Humans are also aware of their own mortality and understand that they have a limited amount of time on planet Earth. Most humans have a desire to be remembered favorably after they have died. The distant future, including the opinions of future generations, matters to them, despite knowing that they are mortal. They can distinguish right from wrong.

Humans can design and create things; their mental abilities and natural talents extend far beyond the basic needs of survival and propagation. Many uniquely human abilities (e.g. the use of clothing or the ability to create and use fire) would obviously be of great benefit to animal survival, but not a single animal species has any of those abilities.

One of the most amazing and profound of human attributes is the desire to discover truths about reality by asking questions about the world around them. Like other living things, humans observe and interact with what they find within their environs. But unlike animals, humans contemplate questions: Who, What, When, Where, Why, and How?

Even chimpanzees, with a genome very close to that of humans, lack every one of the exceptional human mental abilities and natural talents. Why are human beings endowed with so many unique mental abilities and natural talents? Why do none of the animal species possess any of them?

The human mind possesses a unique type of controlled mental processing ability not present in any other living thing. Like a computer system, all forms of life require a physical part and a non-physical software part to function. Operating system software handles basic internal functions; application software handles interactions with the outside world. The application component can include both automatic processes and controlled processes. An automatic life software component works in all living things to ensure their survival. However, humans uniquely possess a controlled life software processing component in the form of controlled mental processes.

Animals lack the controlled mental processes that allow human beings to study things, create things, and to make decisions after contemplating the possible consequences of those decisions. These controlled mental processes explain the vast difference between the behavior of human beings and animals. They also explain the presence of unique human qualities such as being deliberative, poetic, inspired, contemplative, cultural, judgmental, or nostalgic.

Each person has their own unique controlled mental processes, apparent from the huge variation in mental abilities and natural talents. Human beings are far more than just an advanced version of an animal species.

The universe is like a machine, in which every piece must work together correctly or the machine will not work at all. The evidence reveals a universe where physical laws and properties amazingly function in a way that supports stable and persisting forms of matter, energy, and even life.

After considering the evidence and all possible explanations, there remain only two possibilities. Either the universe is the result of a deliberate and intentional plan, or it came from a spontaneous creation event that originated from nothing, yet somehow developed over time into the stable and long-lasting reality that we observe now.

Scientists admit that the spontaneous creation idea is only plausible if a nearly infinite number of universes exist, and they have failed in their attempts to detect any evidence of these other universes. Which is more reasonable, that there are a nearly infinite number of vastly different but undetectable universes that spontaneously appeared from nothing, or that the precisely tuned physical laws and properties of the universe are the result of an intentional plan? The evidence points to the latter.

Some of the most educated people in the world refuse to accept the scientific data and other evidence that points to the intentionally planned universe. Some still insist that the universe must have somehow come about through pure chance. But where is their evidence? The truth is that there is none. How many impossible coincidences will it take to convince some naysayers of the truth? How many fortuitous apparent “accidents" of physics, astronomy, chemistry, and living things does it take? Despite the abundance of evidence in favor of intentional planning, many still hold on to the unsupported assumption that there is no universal plan.

For many decades, numerous scientists have attempted to develop an idea or formulate a mathematical model to permit a "no plan" version of reality; the only result has been the concept of a multiverse spontaneously appearing from nothing.

The multiverse idea requires that each randomly appearing universe must somehow be undetectable from our own universe, since there is no experimental or observational evidence for their existence. This nearly infinite number of universes must have widely varying sets of physical properties so that at least some would be stable, long-lasting, and have properties necessary to support life. And at least one of these universes must be able to produce living things from inanimate matter through some unknown means. Most of these universes, having appeared from nothing, would be unstable and would eventually collapse, vanishing back into nothingness.

Rather than an evidence-based explanation for the nature and origin of the universe, the multiverse proposal looks more like an imaginative storyline for a science fiction novel. If the great English writer, H.G. Wells (1866–1946), were still alive today he could write a best-selling novel from these imaginative ideas to rival his other science fiction stories *The Time Machine, The Invisible Man,* and *The War of the Worlds.*

From Clarity to Resolve

For many, the precise nature of the physical laws and properties of the universe provides sufficient evidence to conclude that the universe has been intentionally planned.

However, additional information independently confirms the reality of the planned universe. Unique human qualities and controlled mental processes, not seen in other forms of life, exist because of special human life software. Moreover, the lack of any evidence for the only other alternative, namely spontaneous creation of a multiverse, leads to only one reasonable conclusion. Intentional planning is the only viable explanation for the nature and origin of our world and the universe.

Despite considerable evidence, there will be some who have difficulty accepting this conclusion. In nearly every case, this is due to one or more of the following three aspects of human nature.

- Taking the Path of Least Difficulty
  Evaluating evidence is not fast or easy. Little time and effort is needed to simply maintain a present point of view, or accept what others claim to be true.

- Social Influence and Peer Pressure
  There is a powerful desire in human nature to be a member of the majority. People will sometimes continue to believe an assertion or theory, despite

contrary evidence, when supported by a community of like-minded individuals. Social pressure from colleagues in an academic field can be a major obstacle to accepting ideas not already established in the minds of their peers.

- Holding on to Long-Established Opinions
  Once an opinion has become well established, it is often difficult to change, even when presented with new relevant information. In their book *The Web of Belief*, Willard V. Quine and J.S. Ullian described this well: "The desire to be right and the desire to have been right are two desires, and the sooner we separate them the better off we are. The desire to be right is the thirst for truth. On all counts, both practical and theoretical, there is nothing but good to be said for it. The desire to have been right, on the other hand, is the pride that goeth before a fall. It stands in the way of our seeing we were wrong, and thus blocks the progress of our knowledge [22]."

The examples from history detailed in Chapter Three show that oftentimes scientific progress is possible only by overcoming these hindrances. Nicholas Copernicus, Galileo Galilei, Barry Marshall, William Harvey, Ignaz Semmelweis, William Coley, and Barry Marshall all experienced substantial resistance from experts of the day. However, evidence ultimately revealed the truth. View the evidence while keeping this in mind. Even an inquiring skeptic must be careful to avoid these possible impediments to progress.

The search to find the true nature of our world and the universe has revealed a universal plan. Where there is a plan there is also a purpose, and an intentionally planned universe requires a planner and architect.

The implication of these results is a subject for a different book. It is my hope that the evidence and conclusions presented here will provide the impetus for those who are curious to investigate further. As with other important things in life, it behooves each of us to investigate the evidence, and to explore where it leads and what it might mean to our futures.

Further interesting insight can be found in a book by homicide detective J. Warner Wallace [23]. J. Warner Wallace is a cold-case homicide investigator and a founding member of the Torrance California Police Department Cold-Case Homicide Unit. In the book *Cold-Case Christianity*, he describes how detectives use clues to solve cold cases and presents compelling evidence that many will find enlightening.

Evidence from the well-planned universe reveals that there is far more to our existence on our little blue planet than meets the eye.

Figure 5. Our Place in the Well-Planned Universe

Notes:

[1] U.S. Congress, *Federal Rules of Evidence,* 1975.

[2] S. Hawking and L. Mlodinow, *The Grand Design.* Bantam books, 2010.

[3] M. Rees, *Just Six Numbers: The Deep Forces that Shape the Universe.* Basic Books, 2000.

[4] J. C. Maxwell, "A dynamical theory of the electromagnetic field," London: The Royal Society, Jan. 1, 1865.

[5] National Aeronautics and Space Administration, "NASA Astrobiology Roadmap," Astrobiology Vol. 8, No. 4, Oct. 9, 2008.
Available:
https://nai.nasa.gov/media/roadmap/2003/g1.html
[Accessed January, 2021]

[6] U.S. Geological Survey Water Science School.
Available: https://water.usgs.gov/edu/water-facts.html
[Accessed January, 2021]

[7] B. J. Carr and M. Y. Rees, "The Anthropic Principle and the Structure of the Physical World," Nature, 278, 605-612, April 12, 1979.
Available: http://dx.doi.org/10.1038/278605a0
[Accessed January, 2021]

[8] A. Linde, and V. Vanchurin, “How many universes are in the multiverse?” Journal Phys.Rev.D81:083525, Oct. 9, 2009.
Available: https://arxiv.org/abs/0910.1589
[Accessed January, 2021]

[9] K. C. Cole, *The Hole in the Universe*. Harcourt Inc., 2001.

[10] C. Mora, D. P. Tittensor, S. Adl, A. G. B. Simpson, and B. Worm, (2011) “How Many Species Are There on Earth and in the Ocean?” PLoS Biol 9(8): e1001127, Aug. 23, 2011. Available:
https://doi.org/10.1371/journal.pbio.1001127
[Accessed January, 2021]

[11] J. Randerson, "Crows match great apes in skillful tool use," London: The Guardian, Aug 17, 2007.

[12] J. H. Wimpenny, A. A. S. Weir, L. Clayton, C. Rutz, and A. Kacelnik, “Cognitive Processes Associated with Sequential Tool Use in New Caledonian Crows,” Public Library of Science: 10.1371/journal.pone.0006471, Aug. 5, 2009.

[13] A. Einstein, *Ideas and Opinions*. Translated by S. Bargmann. London: Crown Publishers, 1954.

[14] N. Polavarapu, G. Arora, V. K. Mittal, and J. F. McDonald, “Characterization and potential functional significance of human-chimpanzee large INDEL variation,” Mobile DNA Journal, Oct. 25, 2011.

Available:
https://mobilednajournal.biomedcentral.com/articles/10.1186/1759-8753-2-13
[Accessed January, 2021]

[15] U.S. Environmental Protection Agency, "Electronics Waste Management in the United States Through 2009," EPA 530-R-11-002, May 2011.

[16] C. Egevang, L. J. Stenhouse, R. A. Phillips, A. Petersen, J. W. Fox, and J. R. D. Silk, "Tracking of Arctic terns Sterna paradisaea reveals longest animal migration," Proceedings of the National Academy of Sciences, 107 (5) 2078-2081, Feb. 2, 2010. Available: https://doi.org/10.1073/pnas.0909493107
[Accessed January, 2021]

[17] "A Connectome of the Adult Drosophila Central Brain," Elife DOI: 10.7554/elife.57443, Sept. 6, 2020. Available:
https://europepmc.org/article/MED/32880371
[Accessed January, 2021]

[18] S. Jobs, Apple Worldwide Development conference, keynote address June 6, 2011.

[19] M. Simpson, "A Sister's Eulogy for Steve Jobs," New York Times, Oct. 30, 2011. Available:
http://www.nytimes.com/2011/10/30/opinion/mona-simpsons-eulogy-for-steve-jobs.html?pagewanted=all.
[Accessed January, 2021]

[20] S. S. Williams, "The terrorist inside my husband's brain," American Academy of Neurology, Sept. 26, 2016.

[21] S. Kwak, H. Kim, J. Chey, and Y. Youm, "Feeling How Old I Am: Subjective Age Is Associated With Estimated Brain Age," Frontiers in Aging Neuroscience, June 7, 2018. Available: https://doi.org/10.3389/fnagi.2018.00168 [Accessed January, 2021]

[22] W. V. Quine and J. S. Ullian, *The Web of Belief.* Random House, 1978.

[23] J. W. Wallace, *Cold-Case Christianity.* David C. Cook, 2013.

ACKNOWLEDGEMENTS

I am fortunate to have found an outstanding editor, Dorothy Irwin. With a keen eye for detail, she helped with formatting and editing—far more than expected—vastly improving the clarity and readability of the book.

I also want to express my appreciation for my parents, Robert C. Davis and Sophia B. Davis, for patiently listening to incessant questions from me at times and for encouraging me to pursue my varied interests during childhood and beyond.

ABOUT THE AUTHOR

Kenneth G. Davis was born in Alberta, Canada and moved to the United States as a child. Returning to Canada, he earned a Bachelor of Science degree in physics from the University of Victoria and was awarded Canada's Governor General's Academic medal for achieving the highest academic average in the graduating class. He has worked in the petroleum industry in both Canada and the United States where he applied computer technology to earth science. Now retired, he feels at home in both countries and enjoys traveling between the two.

Made in the USA
Middletown, DE
20 June 2021

42715785R00083